HIGH-FREQUENCY CIRCUIT ENGINEERING

F. Nibler
and coauthors

THE INSTITUTION OF ELECTRICAL ENGINEERS

Other volumes in this series:

IEE CIRCUITS AND SYSTEMS SERIES 6

Series Editors: Dr D. G. Haigh
Dr R. S. Soin

HIGH- FREQUENCY CIRCUIT ENGINEERING

Originally published in German by Expert-Verlag, 1990

English edition published by The Institution of Electrical Engineers, London, United Kingdom

English edition © 1996: The Institution of Electrical Engineers

The Institution of Electrical Engineers,
Michael Faraday House,
Six Hills Way, Stevenage,
Herts. SG1 2AY, United Kingdom

British Library Cataloguing in Publication Data

A CIP catalogue record for this book
is available from the British Library

ISBN 0 85296 801 9

Printed in England by Bookcraft, Bath

Contents

Foreword

The demand for radio-frequency engineering has increased enormously, partly due to the immense growth of wireless communications. This book brings together the theory and practice relevant to the development of these systems. The volume is a translation (from German) of a book jointly published by Expert-Verlag and the Technical Academy of Esslingen, based on a successful course at the academy. As all the references cited in the German edition are in the German language, we have elected to omit them from this edition as they would be of limited use to an English speaking readership.

The material covers both theoretical principles and practical application for frequencies from the low megahertz range and into the microwave frequency band. The book abounds with sample calculations and design examples. Despite the developing key role of monolithic microwave integrated circuits (MMICs), many systems are based on discrete components in the form of microwave integrated circuits (MICs); in focusing mainly (but not entirely) on the MIC, this book forms a useful companion to Volume 6 in the IEE Circuits & Systems series, 'MMIC design', edited by I. D. Robertson.

The book consists of 10 chapters, each contributed by an expert in the particular field. Chapter 1 begins by describing those network parameters that are useful for developing high-frequency circuits. Chapters 2 and 3 deal with the topic of interconnections for high-frequency components and systems, with Chapter 3 focusing on striplines. Chapter 4 presents the general approach to transistor amplifier design, including matching; Chapter 5 expands on this topic to cover FETs as well as bipolar transistors as the active device, including thermal design and architectures for power amplifiers. Chapters 6 and 7 deal with computer-aided-design (CAD), with Chapter 6 including additional material on S-parameters, noise and optimisation, and Chapter 7 giving an overview of a sample of the many available CAD packages. The key topic of performance parameters, such as dynamic range and sensitivity, types of receiver architecture and components for building receivers, comes into Chapter 8 on receiver components and systems. Chapters 9 and 10 deal with the important subjects of measurement errors and noise. Throughout the book there are many sample calculations and design examples, which help to illustrate the techniques and facilitate their application.

David Haigh
London, December 1995

Authors' preface

The development of semiconductor devices for ever-increasing operating frequencies, ever-increasing power levels, major changes in the technology of circuit components, improvements in measuring technology, the use of computers, increasing demands on circuits, and their increasing capabilities and performance, has played a part in bringing decisive changes to the technology of high-frequency circuits.

One particularly amazing aspect of this is the degree to which the use of computers of all kinds has affected circuit technology in all frequency ranges. Without the computer, there would be no high-quality circuit components, no component and circuit measurements, no optimisation or performance improvement, no network analysis or network synthesis, and no maximisation of economic viability.

Any engineer involved in high-frequency technology, whether already engaged in professional life or whether still being trained, must take account of this, and match their knowledge to the new demands which confront us. In other words, an engineer must be familiar with the basic circuits and the auxiliary quantities used to describe them, the new construction components and their uses, and must be in a position to put to use the new techniques of measurement, circuit design, circuit optimisation, circuit simulation, and circuit set-up.

It is in this context that this book, which is based on a series of lectures given at the Technical Academy of Esslingen, will be of help to the engineer engaged in high-frequency technology — less by way of providing recipes and ready-made problem solutions, but rather by introducing processes, the application of which can resolve particular problems. The basic principles of high-frequency engineering and specific examples from the practical world of circuit development and modern measurement technology have been gathered together by the individual authors, all specialists in their particular areas.

On behalf of all the authors

Professor Dr.-Ing. Ferdinand Nibler, Ottobrunn

Contributors

Prof. Dr. Ing. F. Nibler
Universität der Bundeswehr, FB BET
Neubiberg

Dipl.-Ing. (FH) K. Kupfer
Rohde und Schwarz GmbH & Co. KG
Munich

Prof. Dr. Ing. W. Janssen
Fachhochschule München
Munich

Dipl.-Ing. N. Krausse
Telefunken-Systemtechnik
Ulm

Dipl.-Ing. (FH) G. Lang
Wiltron GmbH
Gilching

Prof. P. Pauli
Universität der Bundeswehr, FB BET
Neubiberg

Dipl.-Ing. A. Rupp
Telefunken-Systemtechnik
Ulm

Dipl.-Ing. F. Schmehr
Rosenberger MICRON
Tittmoning

Chapter 1

Network parameters

F. Nibler

1.1 Network parameters

1.1.1 Introduction

Every circuit in electrical engineering can basically be traced back to the diagram shown in Figure 1.1:

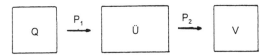

Figure 1.1   *Basic diagram of power transmission system: source* Q *provides power* P_1 *to the transmission element* U, *from which power* P_2 *is passed to the load* V. *All that can be observed is the transmission of energy or power, and not the nature of the transmission, e.g. whether in the form of a 'flow' or a 'wave'*

A source Q feeds a load V via a transmission element U. The power P_1 output from the source Q, and absorbed by the transmission element at the input is, as a rule, different from the power P_2 emitted by the transmission element at the output and absorbed by the load. To describe a system of this kind, three items are required:

- First, an understandable, i.e. easily visualised and at the same time physically correct, *model* of the processes at work in the system
- Secondly, *characteristic values* that are of physical significance and therefore capable of being measured, as well as being suited for the quantitative description of the components
- Thirdly, an *algorithm* to link the characteristic values on the basis of the model.

Electrical engineering knows several models of this type, which can be used from case to case according to suitability or requirements.

1.1.2 LF model and traditional four-terminal network parameters

The most familiar model is probably the *LF model*, which was developed especially for the low frequency range, from which it accordingly takes its name. This model works with the terms J for current and U for voltage, as well as with the two-port network characteristic impedance Z or admittance Y. The transmission element designated as a four-terminal network is described with corresponding parameters.

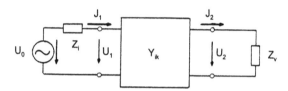

Figure 1.2   *Description of transmission element as four-terminal network, with one pair of input terminals and one pair of output terminals. Transmission characteristics are acquired by a set of four parameters, e.g. by admittance parameters Y_{ik}. Imposed at the terminals are voltages U_1 and U_2, and flowing across the terminals are currents J_1 and J_2. Parameters Y_{ik} can be determined by measuring currents and voltages with a short-circuit at one pair of terminals*

The algorithm that pertains to this provides the theory of linear networks in the form of quadripole (four-pole) equations. With the help of what are known as the admittance parameters Y_{ik}, for example, the equations

$$J_1 \; = \quad Y_{11} \cdot U_1 + Y_{12} \cdot U_2 \tag{1.1a}$$

$$- J_2 \; = \quad Y_{21} \cdot U_1 + Y_{22} \cdot U_2 \tag{1.1b}$$

The values used in this context are readily appreciated, such as 'the input voltage U_1', and can be determined by measurement. Thus, for example, the input impedance Z_e of the quadripole transmission element as shown in Figure 1.2 can be calculated from the measured values 'input voltage U_1' and 'input current J_1' (eqn. 1.2)

$$Z_e \; = \; \frac{U_1}{J_1} \tag{1.2}$$

or the parameter Y_{11} can be determined with port 2 short-circuit, i.e. with $U_2 = 0$, as the 'input short-circuit admittance':

$$Y_{11} \; = \; \frac{J_1}{U_1} \tag{1.3}$$

By analogy, one can also derive the parameter Y_{22} as an 'output short-circuit admittance value', and the parameters Y_{12} and Y_{21} as 'coupling admittance values'.

Entirely by analogy, impedance parameters Z_{ik} can also be introduced for the description of the quadripole, which is considered as a network comprised of discrete elements with current flowing through it; these parameters can be clarified and determined by open-circuit testing, as well as a number of other parameter sets (see also the conversion relationships in Section 1.4).

1.1.3 Line model and wave and chain parameters

The model described in Section 1.1.2 now loses clarity and physical correctness if the transmission element is no longer composed of discrete elements with current flowing through them, and if instead the energy transmission takes place along a line. Calculated and experimental analysis shows that the transport of energy on the line, and in the entire system, then takes place by means of electromagnetic waves. In general, superposition takes place in this context of two waves running in counter directions, usually designated in electrical telecommunications technology as the forward wave and the backward or reflected wave, with opposed energy transport directions. In special cases, only one wave occurs.

The description of the relationships on the special quadripole 'line' is appropriately provided by what are known as 'wave parameters', and by the admittance equations of the quadripole. The *wave parameters* are:

- *Transmission factor* $g = a + jb$ of the line, or the transmission factor $\gamma = \alpha + j\beta = \frac{g}{l}$ related to unit length, with line length l attenuation $a = \alpha \cdot l$, and phase difference $b = \beta \cdot l$

- *Primary and secondary wave impedance* Z_{1w} and Z_{2w}, where, in the case of line transmissions, usually $Z_{1w} = Z_{2w} = Z_w$. The wave impedance is initially derived as a quotient of two current and voltage waves which belong together (i.e. actually H and E-waves), but is then also recognised as the input resistance of an infinitely long line, and can be calculated as a geometrical mean of the open-circuit and short-circuit impedance:

$$Z_w = \sqrt{(Z_0 \cdot Z_k)}$$

The input resistance of a line and of any quadripole is the same as the wave impedance if the load impedance is equal to the secondary wave impedance, i.e. if the line or the quadripole are terminated with the wave impedance*. The line equations are then derived in the form

$$U_1 = U_2 \cdot \cosh(g) + J_2 \cdot Z_w \cdot \sinh(g) \tag{1.4a}$$

---

* A detailed development of the wave parameters and the line equation as a solution to the line differential equation cannot be given here for reasons of space. The interested reader is referred to the relevant specialist literature.

$$J_1 \cdot Z_w = U_2 \cdot \sinh(g) + J_2 \cdot Z_w \cdot \cosh(g) \tag{1.4b}$$

for the symmetrical line, or by analogy for the symmetrical quadripole. In the unsymmetrical case, the following applies with the two wave impedances Z_{1w} and Z_{2w}:

$$U_1 = \frac{\sqrt{Z_{1w}}}{\sqrt{Z_{2w}}} \cdot [U_2 \cdot \cosh(g) + J_2 \cdot Z_{2w} \cdot \sinh(g)] \tag{1.5a}$$

$$J_1 = \frac{1}{\sqrt{(Z_{1w} \cdot Z_{2w})}} \cdot [U_2 \cdot \sinh(g) + J_2 \cdot Z_{2w} \cdot \cosh(g)] \tag{1.5b}$$

In the special case of loss-free, attenuation-free or ideal lines and, as an approximation, also for low-loss lines that are short, the simplified relationships apply:

$$U_1 = U_2 \cdot \cos(b) + j \cdot J_2 \cdot Z_w \cdot \sin(b) \tag{1.6a}$$

$$J_1 \cdot Z_w = j \cdot U_2 \cdot \sinh(b) + J_2 \cdot Z_w \cdot \cosh(b) \tag{1.6b}$$

For the situation of secondary termination with the wave impedance, i.e. with $Z_2 = Z_{2w}$,

$$U_2 = J_2 \cdot Z_{2w} \tag{1.7}$$

and from eqn. 1.5a:

$$U_1 = \frac{\sqrt{Z_{1w}}}{\sqrt{Z_{2w}}} \cdot [U_2 \cdot \cosh(g) + U_2 \cdot \sinh(g)] \tag{1.8}$$

$$U_1 = \frac{\sqrt{Z_{1w}}}{\sqrt{Z_{2w}}} \cdot U_2 \cdot e^g \tag{1.9}$$

and, further

$$g = \ln\left(\frac{U_1}{\sqrt{Z_{1w}}} \cdot \frac{\sqrt{Z_{2w}}}{U_2}\right) \tag{1.10}$$

Accordingly, the transmission value $g = \gamma \cdot l$ can in principle be determined.

The remarks made so far apply, strictly speaking, only to real lines and the waves that occur in this instance. In view of the fact that the waves do not occur explicitly in the final equations (eqns. 1.4a and b, 1.5a and b and 1.6a and b), only the input values U_1 and J_1 (introduced originally as integration constants) and the output values U_2 and J_2 (which are just as capable of measurement at any random quadripole as on a line) can be described with this equation. This also means, however, that the model of energy transmission with waves can be transferred to any transmission element, and any transmission element with wave parameters can therefore be so described.

This consideration also works in reverse and allows for the description of a line with traditional quadripole parameters, which the following is intended to explain, without detailed calculation. The two equations (eqns. 1.1a and b) are resolved according to the input variables:

$$U_1 = U_2 \cdot \frac{-Y_{22}}{Y_{21}} + J_2 \cdot \frac{-1}{Y_{21}} \tag{1.11a}$$

$$J_1 = U_2 \cdot \frac{Y_{12} \cdot Y_{21} - Y_{11} \cdot Y_{22}}{Y_{21}} + J_2 \cdot \frac{-Y_{11}}{Y_{21}} \tag{1.11b}$$

Instead of the fractions, new designations are introduced, initially as a formal abbreviation; this is known as the chain parameter A_{ik}, which can be expressed as

$$U_1 = U_2 \cdot A_{11} + J_2 \cdot A_{12} \tag{1.12a}$$

$$J_1 = U_2 \cdot A_{21} + J_2 \cdot A_{22} \tag{1.12b}$$

This produces a parameter or equation system which can be used to advantage with ladder connections of quadripoles (Section 1.3.2). In addition to this, the equation pair (eqns. 1.12a and b) is structured in the same way as the equation pair (eqns. 1.5a and b), so that a simple factor comparison provides the conversion relationships of the wave parameters to the chain parameters, and so to the conductivity parameters, and vice versa. An example of this is

$$A_{11} = \frac{\sqrt{Z_{1w}}}{\sqrt{Z_{2w}}} \cdot \cosh(g) = \frac{-Y_{22}}{Y_{21}} \tag{1.13}$$

The results of the complete conversion process is shown compiled in the Appendix (Section 1.4).

1.1.4 HF model and scattering and transmission parameters

1.1.4.1 Scattering parameters (S-parameters)
The LF model and to a certain extent the line model can be used to describe circuits of all kinds, from direct current to alternating current of such high frequencies that the wavelengths of the electromagnetic waves associated with them are comparable with the geometric dimensions of the circuit to be described or measured. The short-circuit test referred to earlier, and the measurement at the terminals, should

make this clearer, while it should be borne in mind that there is always a certain distance d between the outer terminals and the active core of a transmission element; as a result, short-circuits or measuring devices can never be connected directly to the active core, as Figure 1.3 shows in schematic form.

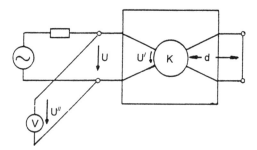

Figure 1.3   *Problems are associated with measurement at the terminals and short-circuit at the terminals if the geometric dimensions of the quadripole are of the same order as the wavelength of the electro-magnetic wave*

- Due to the transformation effect of the lines connected in between, the instrument does not measure the voltage at the terminals, which in turn is not equal to the voltage at the active core.
- Due to the finite distance between the actual active core of the transmission element and its short-circuited terminals, the short-circuit at the terminals is not a short-circuit at the active core, but a load with an unknown complex resistance, which has come about owing to line transformation.

The frequency limit, above which the dimensions cannot as a rule be anything other than small in comparison with the wavelength, lies at about 100 MHz. With higher frequencies, therefore, the LF model, which becomes unusable in this situation, must be replaced by a suitable HF model.

The LF model (and to an extent the line model) is inadequate for another reason; it operates basically with the concept of *current*. The current in this sense flows via conductive connections in a closed circuit, arrives at one terminal of the quadripole in particular, and then leaves it again at the second terminal. However, this means that the LF model (and to an extent the line model) loses any transparency, and its characteristic values become unmeasurable when the energy transmission is effected without conductive connections and without a current circuit, such as via a hollow conductor.

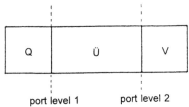

port level 1   port level 2

Figure 1.4   *Transmission element as a two-port network. The source provides power via directly connected transmission element to the load, which in turn is directly connected to the element. At the first transition point, i.e. port level 1, the output port of the source and the input port of the transmission element are in direct connection, and at the second transition point, i.e. port level 2, by analogy, the output port of the transmission element and the input port of the load are in direct connection. The transmission properties are described 'from port to port', incorporating lines inside the transmission element.*

With a model that is based solely on the wave representation, e.g. the *HF model*, which is now to be developed, the relationships can be substantially better described. In this context it is appropriate, instead of the terms 'input' and 'output' with an indication of the terminal pair involved in each case, to use the term 'port', and to do away with the use of the term 'terminal pair', especially in view of the fact that, strictly speaking, it is only the power flow that is observable, and not the current. The description of an arrangement according to Figure 1.4, initially only qualitative, would then read:

The energy-transporting wave leaves the source via the output port, passes into the transmission element through the input port, and passes from the output port of the transmission element into the input port of the load.

Representation using waves provides two immediate advantages: first, connection-free transmissions, such as those in free space or in hollow conductors, can also be described; secondly, by working with waves, one can reliably detect the effects that occur when geometric dimensions and wavelengths are comparable. The range of application of this wave model, or more accurately *HF model*, accordingly ranges up to the highest used frequencies. A more precise observation of the relationships at the transmission two-port network provides the algorithm which pertains to the HF model and the parameters required for the description.

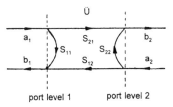

port level 1   port level 2

Figure 1.5   *Signal flow diagram of the transmission element as a two-port network for deriving eqn. 1.14a, b and for definition of the S-parameters*

The significance of the signal flow diagram given in Figure 1.5 can be interpreted as follows: The wave a_1 incoming at the input port gives off part of its energy at the output port, via the internal channel S_{21}, whereupon a wave b_2 leaves the second port; the energy which is not absorbed by the second transmission port leaves the second port via an internal channel S_{11}, as part of the reflected wave b_1 at the input port. By analogy, the wave a_2, incoming at the output, supplies a contribution to b_2 via S_{22}, and to b_1 via S_{12}. This allows the following two equations to be compiled:

$$b_1 = a_1 \cdot S_{11} + a_2 \cdot S_{12} \qquad\qquad (1.14a)$$

$$b_2 = a_1 \cdot S_{21} + a_2 \cdot S_{22} \qquad\qquad (1.14b)$$

This is how the scattering parameters (S-parameters for short) are introduced; the model according to Figure 1.5 is readily described, and the algorithm is provided from eqn. 1.14a, b. What remain to be examined, however, are the physical significance and the measurability of the S-parameters introduced, S_{ik}. The following two postulations serve this purpose, and at the same time represent the basis for specific measurements:

1. The two-port network is connected at the output in such a way that $a_2 = 0$; i.e. either undisturbed radiation of b_2 into space or, more realistically termination with a load Z_B via a line and, if necessary, a measurement arrangement with wave impedance $Z_W = Z_B$, so that no outwardly reflected wave can pass to the output port. The two-terminal network equivalent resistance Z_B of this external consumer, which is entirely independent of the two-port network that is to be measured, is the *reference resistance* for the measurement of the S-parameters. The following then applies:

$$S_{11} = \frac{b_1}{a_1} \text{ and } S_{21} = \frac{b_2}{a_1} \text{ when } a_2 = 0 \qquad\qquad (1.15)$$

It can be seen that

S_{11} is the input S-parameter factor
S_{21} is the forward S-parameter factor

in the circuit described, with the reference resistance Z_B.

2. The two-port network is connected in such a way that at the input $a_1 = 0$; i.e. the power is fed in at port 2, and matches at port 1, as explained. The following then applies:

$$S_{12} = \frac{b_1}{a_2} \text{ and } S_{22} = \frac{b_2}{a_2} \text{ when } a_1 = 0 \qquad\qquad (1.16)$$

It can be seen that

S12 is the reverse S-parameter factor
S22 is the output S-parameter factor

in the circuit described, with the reference resistance Z_B.

It should be emphasised at this point that the freely selectable reference resistance Z_B is incorporated into the measurement. This means that, with one and the same two-port network and different reference resistances, different parameter sets $S_{ik}(Z_B)$ are obtained. In practical HF measuring technology, based on the widespread wave impedance of 50 Ω, the reference resistance chosen almost without exception is $Z_B = 50\ \Omega$, because with this it is easy in most cases to set up a reflection-free measurement arrangement with the existing appliances, lines etc. With 75 Ω devices, the measurement would be just as possible as with a third reference resistance, although in this context three different parameter sets S_{ik} would be derived, all of which would only be usable if the reference resistance were known.

The problem of conversion which arises here is explained in greater detail in Section 1.2.3.

The concept introduced in the preceding very formal manner, of energy-transporting waves a and b with the intensities $|a|^2$ and $|b|^2$, gains substantially in transparency if the HF model is brought into connection with the line model which applies to all frequency ranges; i.e. if the relationships on a line with the current and voltage waves being conducted on it are now described with the aid of the newly introduced a-waves and b-waves, and vice versa the relationships on any transmission two-port network are explained with the aid of current and voltage waves, as in the case of a line. Parameters known from another connection can also be transferred to the transmission two-port network, which is described with S-parameters.

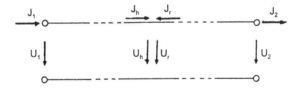

Figure 1.6 *Reference arrows of currents and voltages, or current and voltage waves on lines*

1.1.4.2 Transmission parameters (T-parameters)
The basic relationship for describing the transmission behaviour of a two-port network in the HF model with the aid of the S-parameters is the equation system introduced in Section 1.1.4.1 (eqn. 1.14a and b):

$$b_1 = S_{11} \cdot a_1 + S_{12} \cdot a_2 \tag{1.17a}$$

$$b_2 = S_{21} \cdot a_1 + S_{22} \cdot a_2 \tag{1.17b}$$

This equation system provides the connection between the incoming waves a_1 and a_2 on the one hand, and the reflecting waves b_1 and b_2 on the other, via the S_{ik} of the two-port network in question. Equations 1.17a and b can now be solved with no trouble in accordance with the input values a_1 and b_1, so that

$$b_1 = \frac{-\det(S)}{S_{21}} \cdot a_2 + \frac{S_{11}}{S_{21}} \cdot b_2 \tag{1.18a}$$

$$a_1 = \frac{-S_{22}}{S_{21}} \cdot a_2 + \frac{1}{S_{21}} \cdot b_2 \tag{1.18b}$$

with the abbreviation of the coefficient determinants

$$\det(S) = S_{11} \cdot S_{22} - S_{12} \cdot S_{21} \tag{1.19}$$

In this new equation system, the input values a_1 and b_1 are now shown in their relationship to the output values a_2 and b_2, and it is then logical to introduce four new parameters, known as the *transmission parameters*:

$$T_{11} = \frac{-\det(S)}{S_{21}} \tag{1.20a}$$

$$T_{12} = \frac{S_{11}}{S_{21}} \tag{1.20b}$$

$$T_{21} = \frac{-S_{22}}{S_{21}} \tag{1.20c}$$

$$T_{22} = \frac{1}{S_{21}} \tag{1.20d}$$

This gives the *transmission equations* of the two-port network:

$$b_1 = T_{11} \cdot a_2 + T_{12} \cdot b_2 \tag{1.20a}$$

$$a_1 = T_{21} \cdot a_2 + T_{22} \cdot b_2 \tag{1.20b}$$

The newly introduced parameters can be attributed to the S-parameters with eqns. 1.20a, b, c, and d, and therefore also determine them; they can however also be determined directly by measurement and explained. A distinction is to be drawn between two cases:

1. The two-port network has secondary reflection-free termination, i.e. $a_2 = 0$. In this case

$$T_{12} = \frac{b_1}{b_2} \qquad\qquad T_{22} = \frac{a_1}{b_2}$$

2. The two-port network has a secondary reflection-free feed, i.e. $b_2 = 0$. In this case

$$T_{11} = \frac{b_1}{a_2} \qquad\qquad T_{21} = \frac{a_1}{a_2}$$

In this case, of particular transparency are

- T_{11} as the reverse S-parameter factor
- T_{22} as the forwards S-parameter factor

When measuring the T-parameters according to the layout indicated, it should be borne in mind that, just as with the S-parameters, the external reference resistance Z_B is also incorporated into the result.

With these parameters and eqns. 1.20a and b, the operation, and more especially the ladder network of two two-port networks, can be described particularly easily because the output values of the first two-port network are identical to the input values of the two-port network which follows it.

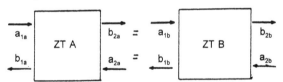

Figure 1.7 *Ladder network of two two-port networks ZT A and ZT B: output values of ZT A are identical to input values ZT B; input values of ZT A are identical to the input values of the cascade, and output values ZT B are identical to output values of the cascade*

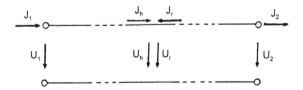

Figure 1.8 *Reference arrows of currents and voltages and current and voltage waves on lines, and in quadripoles and two-port networks which are described by the line model and possibly also by wave parameters*

1.1.5 Reflection factor

At each point x on a line (see Figure 1.8), the outgoing wave (index h, i.e. to the right) and the returning wave (index r, i.e. to the left) are superposed, in such a way that the voltage is

$$U(x) = U_h(x) + U_r(x) \tag{1.21a}$$

and the current

$$J(x) = J_h(x) - J_r(x) \tag{1.21b}$$

The outgoing wave, for example, traverses the line P_h, and has intensity $|a|^2 = P_h$. This power can be attributed formally to the (actual) wave impedance of the line Z_w, the voltage U_h and current J_h:

$$P_h = \frac{U^2_h}{Z_w} = J^2_h \cdot Z_w = |a|^2 \tag{1.22}$$

From this equation,

$$|a| = \sqrt{P_h} \tag{1.23}$$

from which

$$a = \frac{U_h}{\sqrt{Z_w}} \quad \text{or} \quad a = J_h \cdot \sqrt{Z_w} \tag{1.24}$$

with the wave impedance Z_w of the system under consideration, which in this case is regarded as a line for clarity but which does not necessarily have to be a line. By analogy, the following applies to the return wave:

$$b = \frac{U_r}{\sqrt{Z_w}} \quad \text{or} \quad b = J_r \cdot \sqrt{Z_w} \tag{1.25}$$

These can be described jointly as

$$U = U_h + U_r = (a + b) \cdot \frac{1}{\sqrt{Z_w}} \tag{1.26a}$$

$$J = J_h - J_r = (a - b) \cdot \sqrt{Z_w} \tag{1.26b}$$

The quotient

$$Z = \frac{U}{J} = Z_W \cdot \frac{a+b}{a-b} \qquad (1.27)$$

can now be designated as the *two-port equivalent impedance* at the relevant point in the system, and thus also allows for explanations in the wave representation of the HF model with the descriptive terms current, voltage and impedance, even if these do not actually exist in the real physical sense. In particular, this also enables auxiliary energising quantities, parameters, or equivalent circuit diagrams defined elsewhere and in other connections to be transferred to any transmission element and any transmission mode. This can be explained in greater detail with the example of the important reflection factor r.

Reflection factor is defined as

$$r = \frac{U_r}{U_h} = \frac{b}{a} \rightarrow r \cdot a = b \qquad (1.28)$$

From eqns. 1.27 and 1.28, it now follows

$$\frac{Z}{Z_W} = \frac{a+b}{a-b} = \frac{a+r \cdot a}{a-r \cdot a} \qquad (1.29)$$

and further, after eliminating a,

$$\frac{Z}{Z_W} = \frac{1+r}{1-r} \qquad (1.30)$$

From this it further follows, with the two-port equivalent impedance

$$z = \frac{Z}{Z_W} \qquad (1.31)$$

that the reflection factor is

$$r = \frac{z-1}{z+1} \qquad (1.32)$$

Owing to the connections developed here, the relationships in two-port networks and quadripoles, or in one-port and two-port networks, can be described at will, with any model. This means that to describe the relationships at a point inside the system, one can either specify the current, voltage, and impedance at that point, or the reflection factor and the S-parameters as well as the reference impedance belonging to them, and, if necessary, another reference voltage or reference power. By analogy, the transmission behaviour of a two-port network (or quadripole) can be described with any desired parameter system.

A precondition for this, in addition to eqns. 1.30 and 1.32, is a knowledge of the connection between the traditional two-port network parameters (or quadripole parameters) and the S-parameters. This connection is developed in Section 1.1.6.

1.1.6 Conversion calculation for parameters from different models

The transition from LF models with for example Y-parameters to the HF model with S-parameters is obtained by a comparison of the two equation systems with which the two-port network is described:

$$b_1 = S_{11} \cdot a_1 + S_{12} \cdot a_2 \quad (= \text{eqn. 1.14a}) \tag{1.33a}$$

$$b_2 = S_{21} \cdot a_1 + S_{22} \cdot a_2 \quad (= \text{eqn. 1.14b}) \tag{1.33b}$$

$$J_1 = Y_{11} \cdot U_1 + Y_{12} \cdot U_2 \quad (= \text{eqn. 1.1a}) \tag{1.34a}$$

$$-J_1 = Y_{21} \cdot U_1 + Y_{22} \cdot U_2 \quad (= \text{eqn. 1.1b}) \tag{1.34b}$$

In eqn. 1.34 the currents and voltages are now replaced by the corresponding waves according to eqn. 1.26, in which context, to simplify the calculation, symmetry is assumed for the two-port networks under consideration; i.e. $Z_{1W} = Z_{2W} = Z_W$:

$$J_1 = (a_1 - b_1) \cdot \frac{1}{\sqrt{Z_W}} \tag{1.35a}$$

$$J_2 = (a_2 - b_2) \cdot \frac{1}{\sqrt{Z_W}} \tag{1.35b}$$

$$U_1 = (a_1 + b_1) \cdot \sqrt{Z_W} \tag{1.35c}$$

$$U_2 = (a_2 + b_2) \cdot \sqrt{Z_W} \tag{1.35d}$$

Applying eqns. 1.35a – d in eqn. 1.34 leads to

$$\frac{a_1 - b_1}{\sqrt{Z_W}} = Y_{11} \cdot (a_1 + b_1) \cdot \sqrt{Z_W} + Y_{12} \cdot (a_2 + b_2) \cdot \sqrt{Z_W} \tag{1.36a}$$

$$\frac{a_2 - b_2}{\sqrt{Z_W}} = Y_{21} \cdot (a_1 + b_1) \cdot \sqrt{Z_W} + Y_{22} \cdot (a_2 + b_2) \cdot \sqrt{Z_W} \tag{1.36b}$$

With the standardised coefficients

$$y_{ik} = Y_{ik} \cdot Z_W \tag{1.37}$$

then

$$a_1 - b_1 = y_{11} \cdot (a_1 + b_1) + y_{12} \cdot (a_2 + b_2) \tag{1.38a}$$

$$a_2 - b_2 = y_{21} \cdot (a_1 + b_1) + y_{22} \cdot (a_2 + b_2) \tag{1.38b}$$

By converting eqns. 1.38a and b, one obtains

$$-b_1 \cdot (1 + y_{11}) - b_2 \cdot y_{12} = a_1 \cdot (y_{11} - 1) + a_2 \cdot y_{12} \tag{1.39a}$$

$$-b_1 \cdot y_{21} - b_2 \cdot (1 + y_{22}) = a_1 \cdot y_{21} + a_2 \cdot (y_{22} - 1) \tag{1.39b}$$

The two equations can be solved without difficulty for b_1 and b_2:

$$b_1 \cdot [(1 + y_{11}) \cdot (1 + y_{22}) - y_{12} \cdot y_{21}] = \\ a_1 \cdot [(1 - y_{11}) \cdot (1 + y_{22}) + y_{12} \cdot y_{21}] - a_2 \cdot 2 \cdot y_{12} \tag{1.40a}$$

$$b_2 \cdot [(1 + y_{11}) \cdot (1 + y_{22}) - y_{12} \cdot y_{21}] = \\ a_1 \cdot 2 \cdot y_{21} + a_2 \cdot [(1 + y_{11}) \cdot (1 - y_{22}) + y_{12} \cdot y_{21}] \tag{1.40b}$$

A factor comparison in eqns. 1.33 and 1.40 then provides the conversions sought for the transition of the admittance parameters Y_{ik} to the scattering parameters S_{ik}, and therefore simultaneously the transition from the LF model (with current, voltage, and admittance) to the HF model (with waves, wave impedance, reflection factor):

$$S_{11} = \frac{(1 - y_{11}) \cdot (1 + y_{22}) + y_{12} \cdot y_{21}}{(1 + y_{11}) \cdot (1 + y_{22}) - y_{12} \cdot y_{21}} \tag{1.42a}$$

$$S_{11} = \frac{-2 \cdot y_{12}}{(1 + y_{11}) \cdot (1 + y_{22}) - y_{12} \cdot y_{21}} \tag{1.42b}$$

$$S_{21} = \frac{-2 \cdot y_{21}}{(1 + y_{11}) \cdot (1 + y_{22}) - y_{12} \cdot y_{21}} \tag{1.42c}$$

$$S_{22} = \frac{(1 + y_{11}) \cdot (1 - y_{22}) + y_{12} \cdot y_{21}}{(1 + y_{11}) \cdot (1 + y_{22}) - y_{12} \cdot y_{21}} \tag{1.42d}$$

or, with the wave impedance Z_W written explicitly:

$$S_{11} = \frac{(1-Y_{11} \cdot Z_W) \cdot (1+Y_{22} \cdot Z_W) + Y_{12} \cdot Z_W \cdot Y_{21} \cdot Z_W}{(1+Y_{11} \cdot Z_W) \cdot (1+Y_{22} \cdot Z_W) - Y_{12} \cdot Z_W \cdot Y_{21} \cdot Z_W} \qquad (1.43a)$$

$$S_{12} = \frac{-2 \cdot Y_{12} \cdot Z_W}{(1+Y_{11} \cdot Z_W) \cdot (1+Y_{22} \cdot Z_W) - Y_{12} \cdot Z_W \cdot Y_{21} \cdot Z_W} \qquad (1.43b)$$

$$S_{21} = \frac{-2 \cdot Y_{21} \cdot Z_W}{(1+Y_{11} \cdot Z_W) \cdot (1+Y_{22} \cdot Z_W) - Y_{12} \cdot Z_W \cdot Y_{21} \cdot Z_W} \qquad (1.43c)$$

$$S_{22} = \frac{(1+Y_{11} \cdot Z_W) \cdot (1-Y_{22} \cdot Z_W) + Y_{12} \cdot Z_W \cdot Y_{21} \cdot Z_W}{(1+Y_{11} \cdot Z_W) \cdot (1+Y_{22} \cdot Z_W) - Y_{12} \cdot Z_W \cdot Y_{21} \cdot Z_W} \qquad (1.43d)$$

It would be going too far at this point to develop in detail the reversal of this calculation to the transition from S-parameters to admittance parameters, or even to further parameter systems. Some of the most important conversion relationships are therefore compiled in the Appendix, without proof. With the help of these relationships, the transition from one system to another is possible at any time, and, of course, also the transition to the corresponding model and the equivalent circuit on which it is based. This does, however, present the opportunity at any time of transferring, for example, to the descriptive T-element as an equivalent circuit, or of discussing the behaviour of a circuit made of discrete components with the wave representation. This would be important, for example, if two two-port systems were to be connected in series, the properties of which were only known by the S-parameters measured. The transition to the Z-parameters then enables the calculation to be made of the properties of the resultant two-port network by the addition of the two Z-matrices to yield the resultant Z-matrix, from which the transition to the resultant S-parameters in turn presents no problems at all (see also Section 1.3.2).

1.2 Measuring and representing S- and T-parameters

1.2.1 Measuring the parameters

The measurement of the S- or T-parameters can be carried out in a manner suitable for the purpose by means of an arrangement as shown in Figure 1.9. In this context, the waves a_1, a_2, b_1 and b_2 are measured for magnitude and phase in a measuring system with the reference resistance $Z_B = Z_W$, for the various different cases $a_1 = 0$ and $a_2 = 0$, or $b_1 = 0$ and $b_2 = 0$. Because of the need to measure magnitude and phase, the use of measuring devices such as network analysers, or what are known as vector voltmeters, comes into consideration, most of which carry out the actual measurement with computers and the initial evaluation of the measurement on their own, with the result that the complex values of the parameters are directly available at their output as

(1) Measurement of a_1, b_1, b, if $a_2 = 0$

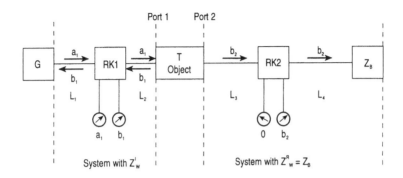

(2) Measurement of a_2, b_1, b, if $a_1 = 0$

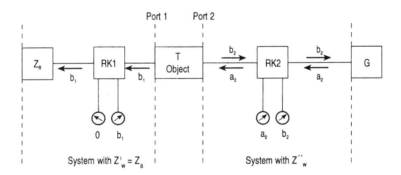

RK   Directional coupler

L_n Lines of any desired length

G   Generator

Z_B Reference resistance

T   Transmission element (measurement object)

Figure 1.9   *Measuring arrangement schematic for measuring S- or T-parameters in two cases*
a) Measurement of a_1, b_1, b_2 *with* $a_2 = 0$
b) Measurement of a_2, b_1, b_2 *with* $a_1 = 0$

$$S_{11}(Z_B) = \frac{b_1}{a_1} \text{ and } S_{21}(Z_B) = \frac{b_2}{a_1} \text{ if } a_2 = 0 \tag{1.44}$$

$$S_{12}(Z_B) = \frac{b_1}{a_2} \text{ and } S_{22}(Z_B) = \frac{b_2}{a_2} \text{ if } a_1 = 0 \tag{1.45}$$

$$T_{11}(Z_B) = \frac{b_1}{a_2} \text{ and } T_{21}(Z_B) = \frac{a_1}{a_2} \text{ if } b_2 = 0 \tag{1.46}$$

$$T_{12}(Z_B) = \frac{b_1}{b_2} \text{ and } T_{22}(Z_B) = \frac{a_1}{b_2} \text{ if } a_2 = 0 \tag{1.47}$$

The T-parameters are also often derived from the S-parameters by internal calculation because in that case just two measurements are sufficient to determine the eight parameters of both systems.

The parameters derived are always dependent on the reference resistance during the measurements, in which context different reference resistances can in principle be used even for the primary and secondary measurements. For obvious reasons, however, in practice the same reference resistance is used in both cases, which is frequently a real resistance of 50 Ω. This reference resistance Z_B is the wave impedance Z_W of the measuring circuit, and must not be confused with the wave impedance Z_{WObj} of the object of which the parameters are being measured. For a symmetrical passive object, for example, the wave impedance is derived from the S-parameters on the basis of the conversion tables as follows:

$$Z_{WObj} = \frac{\sqrt{Z_{11}}}{\sqrt{Y_{11}}} \tag{1.48}$$

$$Z_{WObj} = Z_B \cdot \frac{\sqrt{[(1+S_{11})^2 - S_{12} \cdot S_{21}]}}{\sqrt{[(1-S_{11})^2 - S_{12} \cdot S_{21}]}} \neq Z_B \tag{1.49}$$

when $S_{11} \neq 0$. Only when $S_{11} = 0$ is $Z_{WObj} = Z_B$ also, i.e. in this case the measurement was carried out with the wave impedance of the object as the reference resistance.

1.2.2 Representing the parameters

In view of the fact that when the S- or T-parameters are measured the dependency of the frequency is also acquired, almost without exception, one obtains some comprehensive data for each of the two times four parameters, which is either stored directly in a computer for further processing* there, or must be represented in some other easily visualised manner.

---

* See also Chapters 6 and 7, relating to computer-assisted design of circuits.

The simplest solution is to produce all the values of magnitude and phase, or, for them to be of equal value, to produce all the values of the real and imaginary components in the form of a table. This is, for example, the way in which transistor data are often provided by the manufacturers. An advantage with this is that these values can easily be input into a computer, while a disadvantage is the lack of transparency involved. From this point of view, the diagrammatic representation of the dependency of the complex values of S- and T-parameters on the real parameter, frequency, in the form of a locus diagram is to be preferred.

Two of the four S-parameters, namely S_{11} and S_{22}, have the significance of reflection factors, and are accordingly represented in the reflection factor plane. The other two, namely S_{12} and S_{21}, have the significance of transfer coefficients or actual transformation ratios, and are to be represented in the appropriate planes. For the representation itself, right-angled Cartesian co-ordinates (i.e. real component/imaginary component) or polar co-ordinates (i.e. magnitude/phase) can be used at will or as required, in which context, because of the practical significance of the magnitudes, both the reflection factor and the actual transformation ratio are often preferred to the polar representation. For the circuit shown in Figure 1.10, for example, Figures 1.11a and 1.11b show the locus diagrams of the S-parameters; the nature of the circuit as a bandstop filter is identifiable both from the reflection factor and from the transformation ratio.

Figure 1.12 shows a slightly modified circuit and Figures 1.13a and b the locus diagrams of the S-parameters which pertain to it. By comparison, it can be seen that the circuit shown in Figure 1.10 is plainly the better filter circuit. However, what is intended to be shown here is not this actually rather trivial observation, but rather the possibility of describing the circuit with the locus diagrams of S-parameters. The circuits still retain their properties with the new mode of description; only the type of representation changes.

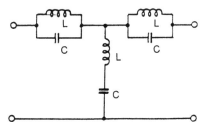

Figure 1.10 *Example representation of S-parameters: loss-free symmetrical LC bandstop filter. Functions in Figures 1.11a, b and c are normalised with $\Omega = \omega \cdot \sqrt{(LC)}$. $Z_B = \sqrt{(L/C)}$ was selected as the reference resistance*

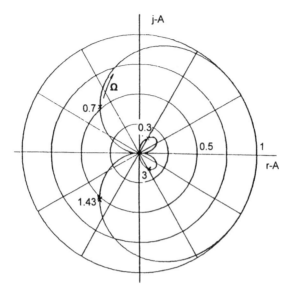

Figure 1.11a *Locus diagram for S$_{11}$ = S$_{22}$ = f(Ω) for bandstop filter shown in Figure 1.10 in polar co-ordinate grid of complex reflection factor plane*

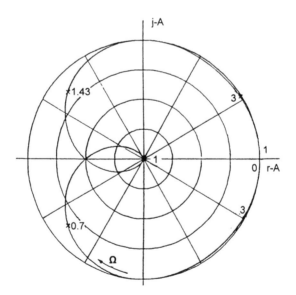

Figure 1.11b *Locus diagram for S$_{12}$ = S$_{21}$ = g(Ω) for bandstop filter shown in Figure 1.10 in polar co-ordinate grid of complex transformation ratio plane*

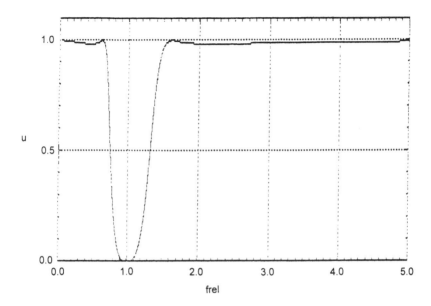

Figure 1.11c *Magnitude of transfer ratio* $|S_{21}|$ *as a function of normalised frequency; filter nature of bandstop filter is particularly apparent*

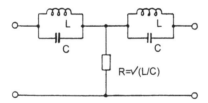

Figure 1.12 *Modified example of representation of S-parameters: imperfect symmetrical bandstop filter. Functions in Figures 1.13a, b and c are normalised with* $\Omega = \omega \cdot \sqrt{(LC)}$, *and* $ZB = R$ *selected as the reference resistance*

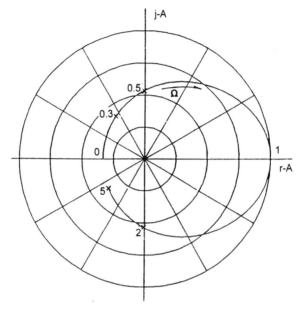

Figure 1.13a *Locus diagram for* $S_{11} = S_{22} = F(\Omega)$ *for bandstop filter shown in Figure 1.12*

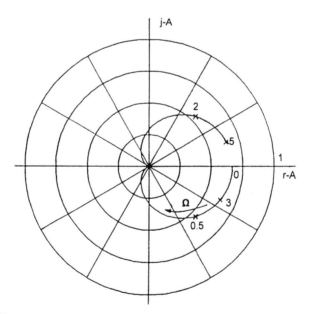

Figure 1.13b *Locus diagram for* $S_{21} = S_{12} = G(\Omega)$ *for bandstop filter shown in Figure 1.12*

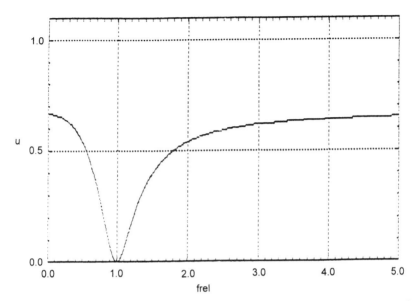

Figure 1.13c *Magnitude of transfer ratio* $|S_{21}|$ *as a function of the normalised frequency. Less satisfactory filter nature of bandstop filter is clearly identifiable in comparison with Figure 1.11c*

What has been shown here for didactic reasons by way of two very simple linear passive networks can also be translated by analogy to larger networks and even to active components or networks. As an expression of the 'amplification', the magnitudes of the transformation ratios and under certain circumstances also of the reflection factors, can become greater than one, i.e. $|S_{ik}| > 1$, and the locus diagrams pertaining to this then also extend into the area outside the unit circle. With passive networks or elements, by contrast, $|S_{ik}| \leq 1$ always applies.

The connection between the reflection factor r, the two-port equivalent impedance Z at a point on the circuit (e.g. the input impedance of a two-port network) and the reference resistance Z_B (e.g. the internal resistance of the source) is provided by the equation

$$r = \frac{Z - Z_B}{Z + Z_B} \tag{1.50}$$

The reflection factor calculated according to this equation has a modulus, in all passive networks, of

$$|r| \leq 1 \tag{1.51}$$

All possible reflection factors r derived from this therefore lie in or on the unit circle of the r-plane. Things are different when we are dealing with active elements, which are to be described formally as negative resistances. In that case, a glance at eqn. 1.50 will show that the modulus of the reflection factor can very well be > 1, which in turn means that the reflection factors belonging to this lie in the r-plane outside the unit circle.

If the reference equation (eqn. 1.50) is used to relate the impedance plane to the reflection factor plane point by point, in other words, to illustrate the impedance plane on the reflection plane with the transforming function eqn. 1.50, one obtains the image points of the passive impedances with the positive effective component in and on the unit circle, outside the image points of the active resistances with the formal negative effective component. The right-hand Z-halfplane, then, corresponds to the inside of the unit circle, and the left-hand Z-halfplane to the outside. The unit circle itself is the image of the imaginary axis of the Z-plane, which delineates the two halfplanes. All impedance values Z are normalised in this context by the reference resistance Z_B. Figure 1.14 shows the results of this representation, namely, the image of the (R,X) grid of the Z-plane in the r-plane with its polar grid. It is clear that, in this case, the range of the unit circle represents the Smith diagram, well known and much used in HF technology.

It can be seen immediately from the Smith diagram, then, which reflection factor r belongs to which (standardised) impedance Z, and vice versa. In particular, the corresponding two-port equivalent impedance Z can be allocated to a specific reflection factor r, and interpreted as an equivalent circuit. These advantages can now be used also in the representation of S_{11} and S_{12}, the locus diagrams of which are entered in the reflection factor plane that supports the Smith diagram. For active configurations, in this context, magnitudes > 1 are possible, and therefore points outside the unit circle. It is absolutely essential to point out this self-evident fact at this point because in practice we use the Smith diagram sheet available, which in the first place comprises only the inside of the unit circle, and, in the second place, for the sake of an easier overview, either does not contain the polar grid of the r-plane at all, or only in a rudimentary manner. This explains the emergence of values outside the diagram, and, on the other hand, also makes clear the lax formulation of the 'entry of S_{11} in the Smith diagram'. An important point in this connection is the possibility of interpreting the S-parameters, or, better, the circuit described by these parameters, by means of the two-port equivalent impedance which can be read from the Smith diagram. One can recognise, for example, from the locus diagram for S_{11} in Figure 1.15, which is identical to the locus diagram shown in Figure 1.13a belonging to the circuit of Figure 1.12, that at low frequencies the input impedance of the circuit is inductive-complex etc., as explained in the caption to Figure 1.15.

The representation of the complex T-parameters as a locus diagram takes place analogously in the corresponding planes, while in that case, however, no combination possibilities or interpretation possibilities exist of the type described.

1.2.3 Effect of reference resistance Z_B

As the S-parameters were introduced, it was pointed out that the reference resistance Z_B is involved in the measurement; in other words, that all S_{ik} are dependent on this

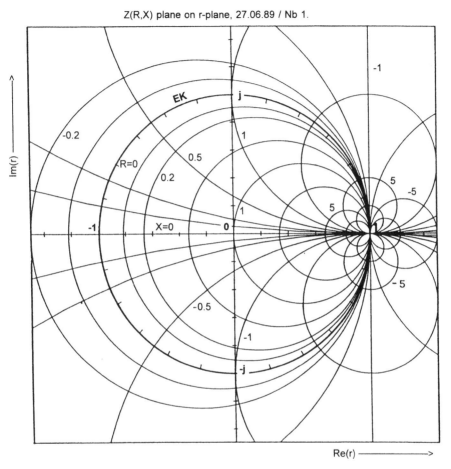

Z(R,X) plane on r-plane, 27.06.89 / Nb 1.

Figure 1.14 *Transforming impedance or Z-plane onto reflection factor or*
r-plane, in accordance with eqn. 1.50. The r-plane is indicated by unit
circle (EK) with 15° marks, as well as by real and imaginary axes; circle
set is image of Cartesian grid of Z-plane. Imaginary axis of Z-plane
(R = 0) is formed on unit circle of r-plane, and real axis of Z-plane
(X = 0) is formed on real axis of r-plane. Figure shows range of Z-plane
– 10 ≤ R ≤ 10 and – 10 ≤ X ≤ 10 by means of grid lines (parameter
values: 0, ± 0.1, ± 0.2, ± 0.5, ± 1, ± 2, ± 5, ± 10). Zero point of Z-plane
is formed at point –1 of r-plane, and infinitely distant point of Z-plane
at point 1 of r-plane. Fixed points of transformation are points ± j.
Part of transformation bordered by unit circle of r-plane contains
image of right-hand Z-halfplane (with positive resistances R ≥ 0),
and forms familiar Smith diagram.

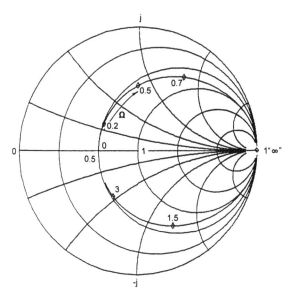

Figure 1.15 *Locus diagram S_{11} (Ω) for circuit from Figure 1.12 entered in Smith diagram. Two-port equivalent resistance for input of circuit loaded with Z_B is, at $\Omega < 1$, identifiably inductive-complex, and at $\Omega > 1$ capacitative-complex. In borderline case $\Omega = 1$, resistance is an infinitely great reactance, with polarity reversal. At very low and very high frequencies, it approximates the real value $0.5 \cdot Z_B$. The equivalent circuit diagram of the input resistance is accordingly a parallel resonance circuit with series resistance (which fully agrees with the actual circuit arrangement).*

value, and that different values for S_{ik} are derived for each two-port network with a different Z_{B2}. This is discussed in greater detail with the example of S_{11}.

The two-port network with the characteristic wave impedance Z_W is loaded secondarily with a load impedance $Z_L = Z_W$, and fed from a source with the internal impedance $Z_i = Z_B$. The primary reflection factor is then

$$r \quad = \quad S_{11} \quad = \frac{Z_W - Z_i}{Z_W + Z_i} \tag{1.52}$$

If, by contrast, a source with the internal impedance $k \cdot Z_i$ is used to supply the feed, the reflection factor derived is

$$r' \quad = \quad \frac{Z_W - k \cdot Z_i}{Z_W + k \cdot Z_i} \neq S_{11} \tag{1.53}$$

From eqn. 1.52 the internal impedance is determined as

$$Z_i = Z_W \cdot \frac{1 - S_{11}}{1 + S_{11}} \tag{1.54}$$

and interpolated in eqn. 1.53, so that

$$r' = \frac{Z_W \cdot (1+S_{11}) - k \cdot Z_W \cdot (1-S_{11})}{Z_W \cdot (1+S_{11}) + k \cdot Z_W \cdot (1-S_{11})} \neq r \tag{1.55}$$

and, after a short transformation,

$$r' = \frac{S_{11} \cdot (1+k) + 1 - k}{S_{11} \cdot (1-k) + 1 + k} \tag{1.56}$$

The new reflection factor r corresponds to a new S_{11} under the new measurement conditions.

The question then arises of how a set $S_{ik}(Z_{B1})$ can be transformed into the new set $S_{ik}(Z_{B2})$ with the new reference resistance Z_{B2}. To answer this question, calculate, for example, from $S_{ik}(Z_{B1})$ and the reference resistance Z_{B1}, the resistance parameters Z_{ik}, in accordance with the conversion relationships in Section 1.4.5. The new S-parameters $S_{ik}(Z_{B2})$ can be calculated from the resistance parameters and the new reference resistance Z_{B2}, in accordance with the conversion relationships of Section 1.4.6.

The following abbreviations are introduced into the calculation:

$$K_{11} = (1+S_{11}) \cdot (1-S_{22}) + S_{12} \cdot S_{21} \tag{1.57a}$$

$$K_{12} = (1-S_{11}) \cdot (1+S_{22}) + S_{12} \cdot S_{21} \tag{1.57b}$$

$$K_{13} = (1-S_{11}) \cdot (1-S_{22}) - S_{12} \cdot S_{21} \tag{1.57c}$$

$$q = \frac{Z_{B2}}{Z_{B1}} \tag{1.57d}$$

According to the relationships of Section 1.4.5,

$$Z_{11} = Z_{B1} \cdot \frac{K_{11}}{K_{13}} \qquad Z_{12} = Z_{B1} \cdot \frac{2 \cdot S_{12}}{K_{13}} \tag{1.58a,b}$$

$$Z_{21} = Z_{B1} \cdot \frac{2 \cdot S_{21}}{K_{13}} \qquad Z_{22} = Z_{B1} \cdot \frac{K_{12}}{K_{13}} \tag{1.58c,d}$$

Equations 1.58 are used in the conversion relationships of Section 1.4.6, so, for example, for S_{11} (Z_{B2}):

$$S_{11}(Z_{B2}) = \frac{(Z_{11} - Z_{B2}) \cdot (Z_{22} + Z_{B2}) - Z_{12} \cdot Z_{21}}{(Z_{11} + Z_{B2}) \cdot (Z_{22} + Z_{B2}) - Z_{12} \cdot Z_{21}} \qquad (1.59)$$

and, after a short manipulation

$$S_{11}(Z_{B2}) = \frac{(K_{11} - q \cdot K_{13}) \cdot (K_{12} + q \cdot K_{13}) - 4 \cdot S_{12} \cdot S_{21}}{(K_{11} + q \cdot K_{13}) \cdot (K_{12} + q \cdot K_{13}) - 4 \cdot S_{12} \cdot S_{21}} \qquad (1.60a)$$

After a longer manipulation, it follows that

$$S_{11}(Z_{B2}) = \frac{(q+1)^2 \cdot S_{11} - (q^2-1) \cdot (1 + \det(S)) + (q-1)^2 \cdot S_{22}}{(q+1)^2 - (q^2-1) \cdot (S_{11} + S_{22}) + (q+1)^2 \cdot \det(S)} \qquad (1.60b)$$

The final result of the calculation, which has only been outlined here and not carried out in full, namely the new set $S_{ik}(Z_{B2})$, is summarised in Section 1.4.13. With the help of this the S-parameters can be converted as required to a new reference resistance (and therefore also the T-parameters).

1.3 Application of parameters

1.3.1 Calculating operational behaviour

1.3.1.1 Preliminary remark
All the examples which follow for the calculation of the operational behaviour are based on a two-port network or quadripole wired as shown in Figure 1.16. The two-port or quadripole is defined in each case in this context by a parameter set P_{ik}.

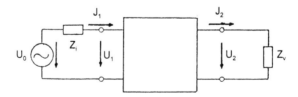

Figure 1.16   *Two-port network (represented as quadripole) connected with source and load components, with reference arrows for currents and voltages*

1.3.1.2 Primary input resistance Z_e

An important value in connection with the feed to the secondarily loaded transmission element is the primary input impedance Z_e. This should be calculated first, for which it is assumed that the S-parameters of the transmission element are known. The direct determination of Z_e from the S-parameters is not possible without further ado; among other things, we do not have the 'comprehensiveness' of the term in the HF model. In such cases, we turn to another, comprehensive, model and convert the result to the parameters desired. The question regarding an impedance is closely associated with the use of impedance parameters (although without this being absolutely mandatory!). With impedance parameters Z_{ik}, the following equations apply:

$$U_1 = J_1 \cdot Z_{11} - J_2 \cdot Z_{12} \tag{1.61a}$$

$$U_2 = J_1 \cdot Z_{21} - J_2 \cdot Z_{22} \tag{1.61b}$$

$$U_2 = J_2 \cdot Z_a \tag{1.62}$$

Combining eqns. 1.61b and 1.62 gives

$$J_2 = J_1 \cdot \frac{Z_{21}}{Z_a + Z_{22}} \tag{1.63}$$

If this expression is used in eqn. 1.61a,

$$Z_e = \frac{U_1}{J_1} = Z_{11} - \frac{Z_{12} \cdot Z_{21}}{Z_a + Z_{22}} \tag{1.64}$$

or, further

$$Z_e = \frac{\det(Z) + Z_{11} \cdot Z_a}{Z_a + Z_{22}} \tag{1.65}$$

The 'correct' (i.e. the desired, or known etc.) parameters are now inserted into this result, which still contains the 'wrong' (i.e. undesired, or unknown) parameters, on the basis of the conversion relationships of Section 1.4.5. After a short manipulation, which is not reproduced here, we obtain

$$Z_e = Z_B \cdot \frac{(Z_B + Z_a) \cdot (1 + S_{11}) + (Z_B - Z_a) \cdot (S_{22} + \det(S))}{(Z_B + Z_a) \cdot (1 - S_{11}) + (Z_B - Z_a) \cdot (S_{22} - \det(S))} \tag{1.66}$$

the final, desired result, namely the input impedance expressed by the S-parameters. The question of understanding of the term 'input impedance' in the wave model does not arise at all; the formal calculation provides the result without any problems, and we could, for example, now connect the transmitter, with appropriate adaptation circuit, directly to the input.

Now look at a particularly interesting special case, namely the case of termination with $Z_a = Z_B$. In this case, the input impedance becomes

$$Z_e = Z_B \cdot \frac{1 + S_{11}}{1 - S_{11}} \tag{1.67}$$

1.3.1.3 Voltage transformation

Another question which frequently arises is that of the *voltage transformation* $u = \dfrac{U_2}{U_1}$. The two-port network is, for example, known by way of its chain parameters A_{ik}. The following equations then apply

$$U_1 \quad = \quad U_2 \cdot A_{11} + J_2 \cdot A_{12} \tag{1.68a}$$

$$J_1 \quad = \quad U_2 \cdot A_{21} + J_2 \cdot A_{22} \tag{1.68b}$$

$$U_2 \quad = \quad J_2 \cdot Z_a \tag{1.69}$$

Combining eqns. 1.69 and 1.68a then immediately provides

$$u \quad = \frac{U_2}{U_1} = \frac{Z_a}{A_{11} \cdot Z_a + A_{12}} \tag{1.70}$$

Assume that the S-parameters are known. By applying the conversion relationships of Section 1.4.7, after a short manipulation

$$u \quad = \frac{2 \cdot S_{21} \cdot Z_a}{Z_a \cdot (1+S_{11}-S_{22}-\det(S)) + Z_B \cdot (1+S_{11}+S_{22}+\det(S))} \tag{1.71}$$

In the special case $Z_a = Z_B$, already discussed, the straightforward expression applies for the voltage transformation according to eqn. 1.72

$$u \quad = \frac{S_{21}}{1 + S_{11}} \tag{1.72}$$

The two-port network should now be connected to a second network in the ladder arrangement (see also Section 1.3.2), so the description with transmission parameters is actually more favourable. The conversion for the special case is effected according to Section 1.4.10, and provides

$$u \quad = \frac{1 \,/\, T_{22}}{1 + T_{12} \,/\, T_{22}} \tag{1.73}$$

$$= \frac{1}{T_{22} + T_{12}} \tag{1.74}$$

1.3.1.4 Relative power transformation p

One very important value, perhaps the most important, of the operational values is the *relative power transformation* p of a two-port network or quadripole; in other words, the active power P in the consuming component relative to the maximum power P_0, which the source could emit. To simplify the calculation, assume that termination is effected with a pure active power-consuming component R_a, and that the source possesses a pure power internal resistance.

The calculation is of course possible with every parameter set. In this instance, purely at random and for demonstration purposes, the set of chain parameters is chosen as the starting point and then converted to find the transmission parameters. The starting point is provided by the chain equations

$$U_1 \quad = \quad U_2 \cdot A_{11} + J_2 \cdot A_{12} \tag{1.75a}$$

$$J_1 \quad = \quad U_2 \cdot A_{21} + J_2 \cdot A_{22} \tag{1.75b}$$

The maximum possible power of the source when $R_i = R_a$ is

$$P_0 \quad = \quad \frac{|U_0|^2}{4 \cdot R_i} \tag{1.76}$$

and the relationships between the secondary values U_2, J_2 and P:

$$U_2 \quad = \quad J_2 \cdot R_a \quad \text{and} \quad P = U_2 \cdot J_2 \rightarrow P = |J_2|^2 \cdot R_a \tag{1.77}$$

From eqns. 1.75a and 1.77,

$$U_1 \quad = \quad J_2 \cdot (A_{11} \cdot R_a + A_{12}) \tag{1.78}$$

and from the eqns. 1.75a and b the input impedance follows:

$$Z_e \quad = \quad \frac{U_1}{J_1} = \frac{A_{11} \cdot R_a + A_{12}}{A_{21} \cdot R_a + A_{22}} \tag{1.79}$$

At the voltage divider, the no-load voltage is divided down to the input voltage from the internal resistance and the input resistance:

$$U_1 = U_0 \cdot \frac{Z_e}{R_i + Z_e} \tag{1.80}$$

Equation 1.80 is used in eqn. 1.79 and again in eqn. 1.78, from which, after a short manipulation, there follows

$$J_2 = U_0 \cdot \frac{1}{A_{11} \cdot R_a + A_{12} + R_i \cdot (A_{21} \cdot R_a + A_{22})} \tag{1.81}$$

Equations 1.76, 1.77 and 1.81 are connected to form

$$p = \frac{P}{P_0} \tag{1.82}$$

$$p = \frac{4 \cdot R_a \cdot R_i}{|A_{11} \cdot R_a + A_{12} + R_i \cdot (A_{21} \cdot R_a + A_{22})|^2} \tag{1.83}$$

$$p = \frac{4 \cdot R_a \cdot R_i}{|A_{11} \cdot R_a + A_{12} + A_{21} \cdot R_i \cdot R_a + A_{22} \cdot R_i|^2} \tag{1.84}$$

In this expression, the relationships according to Section 1.4.12 can be used, and with eqn. 1.85 we obtain

$$p = \frac{4 \cdot R_i \cdot R_a \cdot 4}{\begin{cases} |R_a \cdot (T_{11}+T_{12}+T_{21}+T_{22}) + Z_B \cdot (-T_{11}+T_{12}-T_{21}+T_{22}) \\ + R_a \cdot R_i \cdot Y_B \cdot (-T_{11}+T_{12}+T_{21}+T_{22}) \\ + Z_i \cdot T_{11} - T_{12} - T_{21} + T_{22}|^2 \end{cases}} \tag{1.85}$$

which is an initially confusing and cumbersome expression, but one which would basically already be suitable for a numerical computer evaluation. The systematics of the expression do not become identifiable until some transformations have been effected, and for this reason the following abbreviations are introduced:

$$A_1 = \frac{\sqrt{R_a}}{\sqrt{Z_B}} - \frac{\sqrt{Z_B}}{\sqrt{R_a}} \qquad\qquad A_3 = \frac{\sqrt{Z_B}}{\sqrt{R_i}} - \frac{\sqrt{R_i}}{\sqrt{Z_B}}$$

$$A_2 = \frac{\sqrt{R_a}}{\sqrt{Z_B}} + \frac{\sqrt{Z_B}}{\sqrt{R_a}} \qquad\qquad A_4 = \frac{\sqrt{Z_B}}{\sqrt{R_i}} + \frac{\sqrt{R_i}}{\sqrt{Z_B}}$$

$$p = \frac{4^2}{|(T_{11} \cdot A_1 + T_{12} \cdot A_2) \cdot A_3 + (T_{21} \cdot A_1 + T_{22} \cdot A_2) \cdot A_4|^2} \tag{1.86}$$

A number of special cases are of interest in this instance too:

1. $R_a = Z_B$ leads to $A_1 = 0$ and $A_2 = 2$, and therefore to

$$p_1 \; = \; \frac{4}{|T_{12} \cdot A_3 + T_{22} \cdot A_4|^2} \tag{1.86a}$$

2. $R_i = Z_B$ leads to $A_3 = 0$ and $A_4 = 2$, and therefore to

$$p_2 \; = \; \frac{4}{|T_{21} \cdot A_1 + T_{22} \cdot A_2|^2} \tag{1.86b}$$

3. $R_i = R_a = Z_B$ leads to $A_1 = A_3 = 0$ and $A_2 = A_4 = 2$ and therefore to

$$p_3 \; = \; \frac{1}{|T_{22}|^2} \tag{1.86c}$$

1.3.1.5 Reflection factors r_1 and r_2

The example which follows is intended to show that the behaviour of a two-port network can be calculated with the two *reflection factors* r_1 and r_2, and how this is done. Assume that the two-port network, i.e. a hollow conductor with the characteristic wave impedance Z_W, is known by its S-parameters, and the load at its output by the reflection factor that pertains there, r_2. Which primary reflection factor r_1 is to be anticipated at the input if a generator is connected there with an internal resistance $R_i = Z_W$?

On the output side the following applies:

$$r_2 \; = \; \frac{a_2}{b_2} \tag{1.87}$$

In addition to this, the pair of eqns. 1.14a and b also apply

$$b_1 \; = \; a_1 \cdot S_{11} + a_2 \cdot S_{12} \tag{1.88a}$$

$$b_2 \; = \; a_1 \cdot S_{21} + a_2 \cdot S_{22} \tag{1.88b}$$

Insertion of eqn. 1.87 in eqn. 1.88 provides

$$b_1 \; = \; a_1 \cdot S_{11} + r_2 \cdot b_2 \cdot S_{12} \tag{1.89a}$$

$$b_2 \; = \; a_1 \cdot S_{21} + r_2 \cdot b_2 \cdot S_{22} \tag{1.89b}$$

and from eqn. 1.89b we derive

$$b_2 \quad = \quad \frac{a_1 \cdot S_{21}}{1 - r_2 \cdot S_{22}} \tag{1.90}$$

so that, after insertion of eqn. 1.90 in eqn. 1.89a, the primary reflection factor r_1 is obtained as

$$r_1 \quad = \quad S_{11} + \frac{r_2 \cdot S_{12} \cdot S_{21}}{1 - r_2 \cdot S_{22}} \tag{1.91}$$

or, with the determinant $\det(S)$

$$r_1 \quad = \quad \frac{S_{11} - r_2 \cdot \det(S)}{1 - r_2 \cdot S_{22}} \tag{1.92}$$

An interesting point here is the reversal of the preceding considerations. According to these the reflection factor r_2, which cannot be directly measured because of the inaccessibility of the load, can be calculated from the measurable input reflection factor. From eqn. 1.92 by reversal

$$r_2 \quad = \quad \frac{S_{11} - r_1}{\det(S) - r_1 \cdot S_{22}} \tag{1.93}$$

and with the definition of the reflection factor according to eqn. 1.30, the unknown load impedance is derived as

$$Z \quad = \quad Z_w \cdot \frac{1 + r_2}{1 - r_2} \tag{1.94}$$

$$Z \quad = \quad Z_w \cdot \frac{\det(S) - r_1 \cdot (1 + S_{22}) + S_{11}}{\det(S) + r_1 \cdot (1 - S_{22}) - S_{11}} \tag{1.95}$$

1.3.2 Interconnection of two-port networks or quadripoles

The different circuits in telecommunications engineering are made up from individual elements or modules or, in the case of large-scale circuits such as transmitter systems, from independent devices, all of which can be interpreted as *two-port networks* (or possibly also as multiport networks). In general, it can be said that the (linear) networks in telecommunications engineering are derived from the combination of (linear) two-port networks. The overall characteristics and properties of a network are in this context determined by the characteristics and properties of the individual two-port networks and from the way in which they are combined to form the overall network. The calculation of the transmission characteristics of a network, from the characteristics of its elements, therefore leads initially to the question of how the characteristics of interconnected two-port networks can be determined from the properties of the individual two-port networks.

The transmission characteristics of the individual two-port network are acquired from its transmission parameters, e.g. from the admittance parameters Y_{ik}, which should be known from measurement. This does not however present any kind of restriction, because all parameter sets can be converted from one to another, and therefore if one parameter set is known originally, any desired parameter set can be accessed at any time. The only precondition, then, is that a set of two-port network parameters P_{ik} be known in each case for both the two-port networks to be combined.

There are several possible cases for the combination of two-port networks:

(a) The inputs and outputs are connected in *series*; i.e. the input currents of the interconnected two-port networks are equal to one another, as are the output currents.

(b) The inputs and outputs are connected in *parallel*; i.e. the input voltages are the same, and so are the output voltages.

(c) The two-port networks are connected in *ladder* or *iterative* format; i.e. the output values of the first two-port network are identical to the input values of the second, following, two-port network.

(d) The inputs are connected in *series* and the outputs are connected in *parallel*.

(e) The inputs are connected in *parallel* and the outputs in *series*.

To determine how the properties of the combination can be determined from the properties of its parts, consider the example of the important case of the ladder circuit.

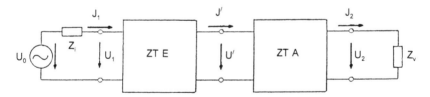

Figure 1.17   *Ladder circuit of two two-port networks (represented as a quadri-pole). Input values of first two-port network ZT E are equal to input values [SiE] of ladder arrangement. Output values of first two-port network are equal to input values of second two-port network ZT A, and output values of second two-port network are equal to output values [SiA] of ladder.*

For the ladder circuit, the description of the two-port network is especially good and clearly identifiable, with the ladder parameters and the equations pertaining to it:

$$U_1 \quad = \quad A_{11} \cdot U_2 + A_{12} \cdot J_2 \tag{1.96a}$$

$$J_1 \quad = \quad A_{21} \cdot U_2 + A_{22} \cdot J_2 \qquad\qquad (1.96b)$$

In the telecommunications engineering sense, the two input values together form the signal, so a symbol compilation to form one single value, the input signal, is logical. To achieve this, form the *signal matrix* [Si_E] of the input signal with the elements U_1 and J_1, and the *signal matrix* [Si_A] of the output signal with the elements U_2 and J_2:

$$[Si_E] = \quad (U_1, J_1) \text{ as a column matrix} \qquad\qquad (1.97)$$

$$[Si_A] = \quad (U_2, J_2) \text{ as a row matrix} \qquad\qquad (1.98)$$

Further, the coefficients A_{ik} of the ladder equations (eqns. 1.96a and b) are compiled to form the *iterative matrix* [A], or the matrix of the ladder network

$$[A] \quad = \begin{bmatrix} A_{11} & A_{12} \\ A_{21} & A_{22} \end{bmatrix} \qquad\qquad (1.99)$$

With these matrices, eqns. 1.96a and b can be written as

$$[Si_E] \quad = [A] \cdot [Si_A] \qquad\qquad (1.100)$$

In particular, for the two-port network 1

$$[Si_E]_1 \quad = [A]_1 \cdot [Si_A]_1 \qquad\qquad (1.101)$$

and for the two-port network 2

$$[Si_E]_2 \quad = [A]_2 \cdot [Si_A]_2 \qquad\qquad (1.102)$$

From Figure 1.17 it can now be clearly seen that

• the output signal of the first two-port network is the same as the input signal of the second two-port network

$$[Si_A]_1 \quad = [Si_E]_2$$

• the input signal of the first two-port network is the same as the input signal of the ladder arrangement

$$[Si_E]_K \quad = [Si_E]_1$$

• the output signal of the second two-port network is the same as the output signal of the ladder arrangement

$[S_{iA}]_K = [S_{iA}]_2$

This allows eqns. 1.101 and 1.102 to be combined to

$$[S_{iE}]_1 = [A]_1 \cdot [A]_2 \cdot [S_{iA}]_2 \tag{1.103}$$

With the corresponding matrix $[A]_K$ of the ladder network, the new two-port network formed by the ladder connection of the two parts, the following can undoubtedly apply:

$$[S_{iE}]_K = [A]_K \cdot [S_{iA}]_K \tag{1.104}$$

Taking into account the equality of the corresponding signals, as described above, a comparison of Equations 1.103 and 1.104 produces the relationship which is being sought:

$$[A]_K = [A]_1 \cdot [A]_2 \tag{1.105}$$

This means, then, that the properties of the resultant two-port network can be calculated from the properties of the two combined two-port networks, in the case of the ladder network by the multiplication of the iterative, or ladder, matrices. The individual element P_{ik} of a product matrix $[P]$ is calculated in this context from the elements Q_{ik} and R_{ik} of the factor matrices $[Q]$ and $[R]$ according to eqn. 1.106

$$[P] = [Q] \cdot [R] \tag{1.106a}$$

$$P_{ik} = \sum_{m=1}^{m=2} (Q_{im} \cdot R_{mk}) \tag{1.106b}$$

in the special case of quadratic matrices with 2×2 elements, which are particularly important for telecommunications engineering.

For the case of the combination of several two-port networks, to form one n-element ladder, one can immediately generalise

$$[A]_K = \prod_{m=1}^{m=n} ([A]_m) \tag{1.107}$$

The calculation of the ladder circuit is also possible with no trouble by using transmission parameters, since in this case there applies, with the corresponding matrices $[T]_1$, $[T]_2$ and $[T]_K$

$$[T]_K = [T]_1 \cdot [T]_2 \tag{1.108}$$

and for an n-element ladder

$$[T]_K = \prod_{m=1}^{m=n} ([T]_m) \tag{1.109}$$

To calculate the properties of a combination in the other cases outlined above, the following matrices are initially formed by analogy to eqn. 1.99

[Z]   from the impedance parameters
[Y]   from the admittance parameters
[H]   from the H-parameters
[G]   from the G-parameters

The calculation is then carried out as follows (without proof):

(a) *Series connection* of n two-port networks

$$[Z] = \sum_{m=1}^{m=n} ([Z]_m)$$

(b) *Parallel connection* of n two-port networks

$$[Y] = \sum_{m=1}^{m=n} ([Y]_m)$$

(d) *Series-parallel connection* of n two-port networks

$$[H] = \sum_{m=1}^{m=n} ([H]_m)$$

(e) *Parallel-series connection* of n two-port networks

$$[G] = \sum_{m=1}^{m=n} ([G]_m)$$

Note to (d) and (e): The G- and H-parameters are not presented in detail in this book, and are mentioned here only for reasons of completeness. The specialist literature should be consulted for further information.

It was assumed at the outset that the conductivity parameters of the two two-port networks interconnected in the ladder were known. If we further assume that, in the final analysis, the S-parameters of the resultant ladder are needed for further work, then the following procedure is derived (which would need to be modified accordingly in other combination forms):

- Selection of the appropriate parameters and the corresponding conversion, in this case $Y_{ik} \rightarrow A_{ik}$, in accordance with Section 1.4.1 in the Appendix.
- Calculation of the combination circuit, in this case $[A]_K = [A]_1 \cdot [A]_2$.
- Conversion of the parameters determined into the parameters which are finally being sought, in this case $A_{ik} \rightarrow S_{ik}$, in accordance with Section 1.4.8 in the Appendix.

The reader can find a continuation and application of the ideas developed here in Chapter 6.

1.4 Appendix: conversion tables

Abbreviations

Reference resistance	Z_B
Admittance parameters	Y_{ik}
Impedance parameters	Z_{ik}
Ladder parameters	A_{ik}
Scatter or S-parameters	S_{ik}
Transmission or T-parameters	T_{ik}
Transmission factor	$g = a + jb$

$Y_B = 1/Z_B$

$det(P) = P_{11} \cdot P_{22} - P_{12} \cdot P_{21}$

1.4.1 Conversion $A_{ik} = f(Y_{ik})$

$$A_{11} = \frac{-Y_{22}}{Y_{21}} \qquad A_{12} = \frac{-1}{Y_{21}}$$

$$A_{21} = \frac{-det(Y)}{Y_{21}} \qquad A_{22} = \frac{-Y_{11}}{Y_{21}}$$

1.4.2 Conversion $Y_{ik} = f(A_{ik})$

$$Y_{11} = \frac{A_{11}}{A_{12}} \qquad\qquad Y_{12} = \frac{-\det(A)}{A_{12}}$$

$$Y_{21} = \frac{1}{A_{12}} \qquad\qquad Y_{22} = \frac{-A_{11}}{A_{12}}$$

1.4.3 Conversion $Y_{ik} = f(S_{ik})$

$$Y_{11} \cdot Z_B = \frac{(1+S_{22}) \cdot (1-S_{11}) + S_{12} \cdot S_{21}}{(1+S_{11}) \cdot (1+S_{22}) - S_{12} \cdot S_{21}}$$

$$Y_{12} \cdot Z_B = \frac{-2 \cdot S_{12}}{(1+S_{11}) \cdot (1+S_{22}) - S_{12} \cdot S_{21}}$$

$$Y_{21} \cdot Z_B = \frac{-2 \cdot S_{21}}{(1+S_{11}) \cdot (1+S_{22}) - S_{12} \cdot S_{21}}$$

$$Y_{22} \cdot Z_B = \frac{(1+S_{11}) \cdot (1-S_{22}) + S_{12} \cdot S_{21}}{(1+S_{11}) \cdot (1+S_{22}) - S_{12} \cdot S_{21}}$$

1.4.4 Conversion $S_{ik} = f(Y_{ik})$

$$S_{11} = \frac{(1-Y_{11} \cdot Z_B) \cdot (1+Y_{22} \cdot Z_B) + Y_{12} \cdot Z_B \cdot Y_{21} \cdot Z_B}{(1+Y_{11} \cdot Z_B) \cdot (1+Y_{22} \cdot Z_B) - Y_{12} \cdot Z_B \cdot Y_{21} \cdot Z_B}$$

$$S_{12} = \frac{-2 \cdot Y_{12} \cdot Z_B}{(1+Y_{11} \cdot Z_B) \cdot (1+Y_{22} \cdot Z_B) - Y_{12} \cdot Z_B \cdot Y_{21} \cdot Z_B}$$

$$S_{21} = \frac{-2 \cdot Y_{21} \cdot Z_B}{(1+Y_{11} \cdot Z_B) \cdot (1+Y_{22} \cdot Z_B) - Y_{12} \cdot Z_B \cdot Y_{21} \cdot Z_B}$$

$$S_{22} = \frac{(1+Y_{11} \cdot Z_B) \cdot (1-Y_{22} \cdot Z_B) + Y_{12} \cdot Z_B \cdot Y_{21} \cdot Z_B}{(1+Y_{11} \cdot Z_B) \cdot (1+Y_{22} \cdot Z_B) - Y_{12} \cdot Z_B \cdot Y_{21} \cdot Z_B}$$

1.4.5 Conversion $Z_{ik} = f(S_{ik})$

$$Z_{11} = Z_B \cdot \frac{(1+S_{11}) \cdot (1-S_{22}) + S_{12} \cdot S_{21}}{(1-S_{11}) \cdot (1-S_{22}) - S_{12} \cdot S_{21}}$$

$$Z_{12} = Z_B \cdot \frac{2 \cdot S_{12}}{(1-S_{11}) \cdot (1-S_{22}) - S_{12} \cdot S_{21}}$$

$$Z_{21} = Z_B \cdot \frac{2 \cdot S_{21}}{(1-S_{11}) \cdot (1-S_{22}) - S_{12} \cdot S_{21}}$$

$$Z_{22} = Z_B \cdot \frac{(1-S_{11}) \cdot (1+S_{22}) + S_{12} \cdot S_{21}}{(1-S_{11}) \cdot (1-S_{22}) - S_{12} \cdot S_{21}}$$

1.4.6 Conversion $S_{ik} = f(Z_{ik})$

$$S_{11} = \frac{(Z_{11}-Z_B) \cdot (Z_{22}+Z_B) - Z_{12} \cdot Z_{21}}{(Z_{11}+Z_B) \cdot (Z_{22}+Z_B) - Z_{12} \cdot Z_{21}}$$

$$S_{12} = \frac{Z_B \cdot 2 \cdot Z_{12}}{(Z_{11}+Z_B) \cdot (Z_{22}+Z_B) - Z_{12} \cdot Z_{21}}$$

$$S_{21} = \frac{Z_B \cdot 2 \cdot Z_{21}}{(Z_{11}+Z_B) \cdot (Z_{22}+Z_B) - Z_{12} \cdot Z_{21}}$$

$$S_{22} = \frac{(Z_{11}+Z_B) \cdot (Z_{22}-Z_B) - Z_{12} \cdot Z_{21}}{(Z_{11}+Z_B) \cdot (Z_{22}+Z_B) - Z_{12} \cdot Z_{21}}$$

1.4.7 Conversion $A_{ik} = f(S_{ik})$

$$A_{11} = \frac{(1+S_{11}) \cdot (1-S_{22}) + S_{12} \cdot S_{21}}{2 \cdot S_{21}}$$

$$A_{12} = \frac{Z_B \cdot [(1+S_{11}) \cdot (1+S_{22}) - S_{12} \cdot S_{21}]}{2 \cdot S_{21}}$$

$$A_{21} = \frac{(1-S_{11}) \cdot (1-S_{22}) - S_{12} \cdot S_{21}}{Z_B \cdot 2 \cdot S_{21}}$$

$$A_{22} = \frac{(1-S_{11}) \cdot (1+S_{22}) + S_{12} \cdot S_{21}}{2 \cdot S_{21}}$$

1.4.8   Conversion $S_{ik} = f(A_{ik})$

$$S_{11} = \frac{(A_{11}-A_{22}) - Z_B \cdot A_{21} + Y_B \cdot A_{12}}{(A_{11}+A_{22}) + Z_B \cdot A_{21} + Y_B \cdot A_{12}}$$

$$S_{12} = \frac{2 \cdot \det(A)}{(A_{11}+A_{22}) + Z_B \cdot A_{21} + Y_B \cdot A_{12}}$$

$$S_{21} = \frac{2}{(A_{11}+A_{22}) + Z_B \cdot A_{21} + Y_B \cdot A_{12}}$$

$$S_{22} = \frac{(A_{22}-A_{11}) - Z_B \cdot A_{21} + Y_B \cdot A_{12}}{(A_{11}+A_{22}) + Z_B \cdot A_{21} + Y_B \cdot A_{12}}$$

1.4.9   Conversion $T_{ik} = f(S_{ik})$

$T_{11} = -\det(S)/S_{21}$

$T_{12} = S_{11}/S_{21}$

$T_{21} = -S_{22}/S_{21}$

$T_{22} = 1/S_{21}$

1.4.10   Conversion $S_{ik} = f(T_{ik})$

$S_{11} = T_{12}/T_{22}$

$S_{12} = \det(T)/T_{22}$

$S_{21} = 1/T_{22}$

$S_{22} = -T_{12}/T_{22}$

1.4.11   Conversion $T_{ik} = f(A_{ik})$

$T_{11} = 0.5 \cdot [A_{11} + A_{22} - Z_B \cdot A_{21} - Y_B \cdot A_{12}]$

$T_{12} = 0.5 \cdot [A_{11} - A_{22} - Z_B \cdot A_{21} + Y_B \cdot A_{12}]$

$T_{21} = 0.5 \cdot [A_{11} - A_{22} + Z_B \cdot A_{21} - Y_B \cdot A_{12}]$

$T_{22} = 0.5 \cdot [A_{11} + A_{22} + Z_B \cdot A_{21} + Y_B \cdot A_{12}]$

1.4.12   Conversion $A_{ik} + f(T_{ik})$

$A_{11} = 0.5 \cdot (T_{11} + T_{12} + T_{21} + T_{22})$

$A_{12} = 0.5 \cdot Z_B \cdot (- T_{11} + T_{12} - T_{21} + T_{22})$

$A_{21} = 0.5 \cdot Y_B \cdot (- T_{11} - T_{12} + T_{21} + T_{22})$

$A_{22} = 0.5 \cdot (T_{11} - T_{12} - T_{21} + T_{22})$

1.4.13   Conversion $S_{ik-1}(Z_{B1}) \rightarrow S_{ik-2}(Z_{B2})$

$Z_{B2} = q \cdot Z_{B1}$

$$S_{1+2} = \frac{(q+1)^2 \cdot S_{11-1} - (q^2-1) \cdot (1+\det(S_1)) + (q-1)^2 \cdot S_{22-1}}{(q+1)^2 - (q^2-1) \cdot (S_{11-1}+S_{22-1}) + (q-1)^2 \cdot \det(S_1)}$$

$$S_{12-2} = \frac{S_{12-1} \cdot 4 \cdot q}{(q+1)^2 - (q^2-1) \cdot (S_{11-1}+S_{22-1}) + (q-1)^2 \cdot \det(S_1)}$$

$$S_{21-2} = \frac{S_{21-1} \cdot 4 \cdot q}{(q+1)^2 - (q^2-1) \cdot (S_{11-1}+S_{22-1}) + (q-1)^2 \cdot \det(S_1)}$$

$$S_{22+2} = \frac{(q+1)^2 \cdot S_{22-1} - (q^2-1) \cdot (1+\det(S_1)) + (q-1)^2 \cdot S_{11-1}}{(q+1)^2 - (q^2-1) \cdot (S_{11-1}+S_{22-1}) + (q-1)^2 \cdot \det(S_1)}$$

1.4.14   Conversion $A_{ik} = f(Z_{iw},g)$

$A_{11} = \dfrac{\sqrt{Z_{1w}}}{\sqrt{Z_{2w}}} \cdot \cosh(g)$

$A_{12} = \sqrt{(Z_{1w} \cdot Z_{2w})} \cdot \sinh(g)$

$A_{21} = \dfrac{1}{\sqrt{(Z_{1w} \cdot Z_{2w})}} \cdot \sinh(g)$

$A_{22} = \dfrac{\sqrt{Z_{w2}}}{\sqrt{Z_{w1}}} \cdot \cosh(g)$

1.4.15 det(Z)

$$\det(Z) = Z_B^2 \cdot \frac{1 + S_{11} + S_{22} + \det(S)}{1 - S_{11} - S_{22} + \det(S)}$$

1.4.16 det(S)

$$\det(S) = \frac{Z_B \cdot (Z_B - Z_{11} - Z_{22}) + \det(Z)}{Z_B \cdot (Z_B + Z_{11} + Z_{22}) + \det(Z)}$$

Chapter 2

Transmission media

W. Janssen

2.1 General line selection

The lines of electromagnetic waves are important components in high frequency engineering. A determinant requirement for lines is the point-to-point transfer without any notable loss of radiant heat and ohmic loss (I^2R loss). The type of line in this situation often characterises the circuit engineering techniques used in a particular frequency range.

A distinction is drawn essentially between three typical types of line:

— Lecher lines or TEM lines (two-wire lines, coaxial lines, striplines)
— hollow conductors (rectangular or circular hollow conductors)
— surface lines (dielectric lines)

In the case of Lecher lines, the electrical and magnetic fields are polarised vertically to the direction of propagation. In the case of hollow lines, electrical and magnetic field components are also in the direction of propagation; the case with surface lines is similar.

Figure 2.1 shows some of the most important types of line. In this context, the typical frequency ranges, attenuations, transferable continuous power, and different scopes of application are shown allocated to the different types of line. Important considerations in selecting the type of line for a given frequency range are: good attenuation values, transferable power, good reproducibility, manufacturing price and expenditure on labour, strength, and the possibility of miniaturisation.

With frequencies of more than about 10 MHz, it is recommended in many cases that a departure be made from the open format or a two-wire line. In such cases, closed-line systems should be used, such as coaxial lines, hollow conductors, and striplines.

The question which arises over and over again for the engineer is what line lengths are permissible in circuits which are composed of dense components. It is often the case that circuit units composed, for example, of integrated modules are connected to one another (Figure 2.2). Connecting lines are frequently flexible strip transmission lines or twisted single-strand bundles. In this context, copper conductors (flexible leads) are embedded in a dielectric compound material. The effective mean

Line type	Structure	Frequency (typical)	Attenuation (typical)	Max. transferable continuous output (typical)	Application
two-wire line	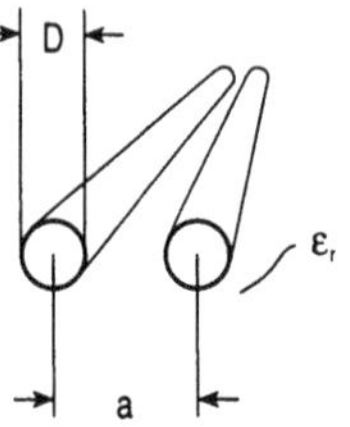 $Z_l/\Omega = \dfrac{120}{\sqrt{\varepsilon_r}} \text{ arcosh } \dfrac{a}{D}$	0 to 100 MHz	0.01 to 1 dB/m	up to 10^9 W	energy transfer, signal transfer, circuits
coaxial line	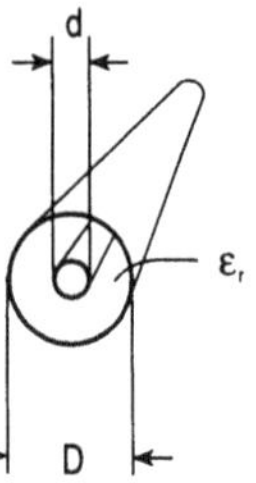 $Z_l/\Omega = \dfrac{60}{\sqrt{\varepsilon_r}} \ln \dfrac{D}{d}$	0 to 20 GHz	0.1 to 10 dB/m	up to 10^7 W	energy transfer, signal transfer, circuits
microstrip line	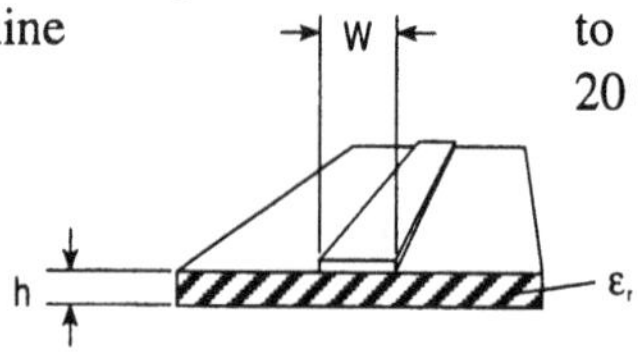 $Z_l/\Omega = F(w, h, \varepsilon)$	100 MHz to 20 GHz	0.1 to 20 dB/m	to 100 W	signal transfer, circuits, hybrid circuits

Figure 2.1 *Line types in high-frequency engineering; structure and typical characteristics*

Line type	Structure	Frequency (typical)	Attenuation (typical)	Max. transferable continuous output (typical)	Application
rectangular hollow conductor	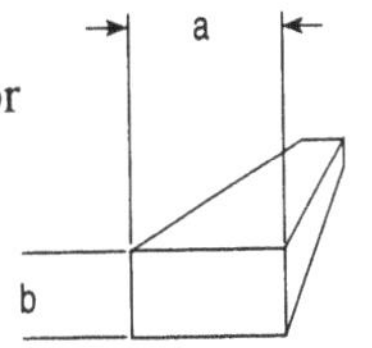	1 to 150 GHz	0.01 to 10 dB/m	400 kW at 1 GHz 30 W at 100 GHz	radar, radio relay, antenna, signal transfer, circuits
circular hollow conductors	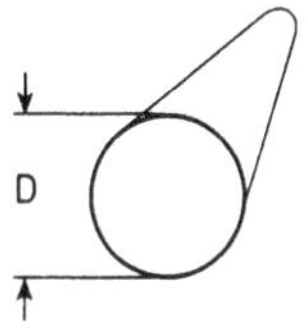				
dielectric lines	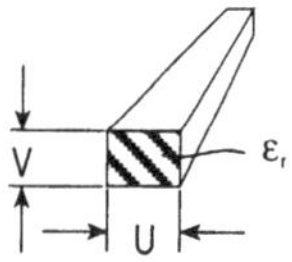	0.1 to 50 GHz	to 10 dB/m	s/kW	antenna, earth-free signal transfer
optical fibres	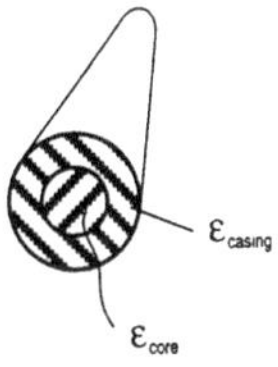	200-600 THz	1 to 10 dB/km	100 mW	signal transfer, earth-free line

For the rectangular hollow conductor:

$$Z_L/\Omega = K \frac{377}{\sqrt{1 - \left(\frac{\lambda_0}{2a}\right)^2}}$$

For the dielectric lines:

$$Z_L/\Omega = F(U, V, \varepsilon_r)$$

Figure 2.1 *Line types in high-frequency engineering; structure and typical characteristics (continued)*

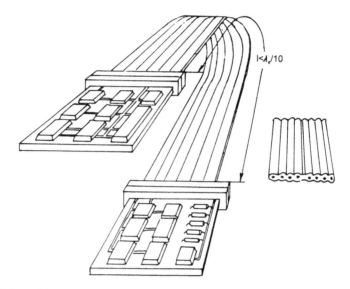

Figure 2.2   *Line as connecting element between circuit units*

dielectric constant is assumed in this case, by way of an example from stripline technology, to be $\varepsilon_r = 2$.

The highest frequency which occurs in signal transfer is 30 MHz; e.g. data processing systems running with a 10 MHz clock frequency. From the value of the highest frequency transferred between the two PCBs illustrated in Figure 2.2, we derive a wavelength in air of $\lambda_0 = c/f$ = speed of light/frequency; $\lambda_0 = 10$ m. Due to the fact that the carrier material incurs a shortening of the wavelength, $\lambda_\varepsilon = \lambda_0/\sqrt{\varepsilon_r} = 7$ m.

Line lengths l are not to be regarded as short in the case of dense circuits until $l < \lambda_\varepsilon/10$. If the line is longer, line theory considerations, such as adaptation, line reflection, etc. must be applied. In the example under consideration, then, the line should not be longer than $l = 70$ cm. However, even in cases in which this condition $l < \lambda_\varepsilon/10$ is fulfilled, difficulties still arise if the principle of coupling-free line guidance is not fulfilled. It is not sufficient for different circuit points simply to be connected purely electrically; account must be taken of the character of the line in the context of electrical transmissions, as otherwise undefined capacitative or inductive circuit couplings will occur. Perfect line routes are automatically created across a common earth by means of double-laminated PCBs; and defined lines can also be represented by means of broad-surface mass structures or direct voltage feeds (which undergo high-frequency short-circuiting to earth) (Figure 2.3).

In the case of striplines, earth connections are usually incorporated. This is how signal lines are evolved that will guarantee interference-free transfer without power-circuit couplings. Even braided-line phases, which are schematically shown to cross one another, can avoid coupling.

If it is intended that lines should be longer than the previously required length $l < \lambda_\varepsilon/10$, then to obtain interference-free transmission the internal resistance of the source R_i (Figure 2.4), the cable wave impedance Z_L, and the terminating resistor R_a must be equal. This basic requirement for an optimum signal transfer applies to all types of lines; in other words, just as much with two-wire lines as with coaxial lines, with striplines, hollow conductors, and dielectric lines. The wave impedances Z_L are shown for some lines in Figure 2.1, provided that they do not result in excessively complicated expressions for striplines and dielectric lines.

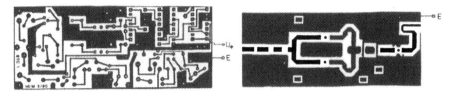

Figure 2.3a *Low-coupling circuit layout on double-laminated PCB*
     b *Stripline circuit set up for correct wave impedance without voltage or current coupling*

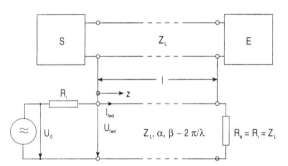

Figure 2.4 *Basic line transmission system*

The surface conductors are of limited significance in high-frequency engineering, being restricted to relatively few areas of application. However, when provided with circular coated cross-sections, surface conductors are being increasingly used in the fibre optic sector as waveguides.

2.2 Wave propagation in lines

2.2.1 Reflection-free transmission

On lines, electromagnetic waves are transferred. These waves are identified by their electric (E) and magnetic (H) fields which have vector properties, and are therefore directional quantities. In the case of lines, electrical fields can be replaced by voltages U, and magnetic fields by current J.

The relationships between electric and magnetic fields, furnished by the basic laws of electrical engineering (Maxwell's equations), can be represented by means of simple relationships between voltages and currents (Figure 2.5).

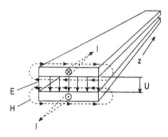

Figure 2.5   *Representation of flat electromagnetic wave as voltage and current waves in parallel plate line*

Voltage and current waves are derived in cases of nonreflection

$$\underline{U}(z) = U_{fwd}\, \varepsilon^{-\alpha z}\, \varepsilon^{-j2\pi z/\lambda} = U_{fwd}\, \varepsilon^{-\gamma z} \qquad (2.1)$$

$$\underline{J}(z) = J_{fwd}\, \varepsilon^{-\alpha z}\, \varepsilon^{-j2\pi z/\lambda} = J_{fwd}\, \varepsilon^{-\gamma z} \qquad (2.2)$$

In eqns. 2.1 and 2.2, $\underline{U}(z)$ and $\underline{J}(z)$ are the complex z-dependent voltages and currents. U_{fwd} and J_{fwd} are the effective values of the voltages and currents running in the positive z-direction (i.e. in the direction of propagation). These values are assumed in this context to be real quantities. This implies that, with $z = 0$, the real power

$$P = U_{fwd} \cdot J_{fwd} \qquad (2.3)$$

is fed in. In general, the lines feature dissipative power losses. These losses are described by the damping constant α, which is usually expressed in dB/m. In eqns. 2.1 and 2.2, Neper per metre must be used for α. The conversion is provided by eqn. 2.4:

$$\alpha[Np/m] = (1/8.7)\, \alpha_{dB}\, [dB/m] \qquad (2.4)$$

The phase constant ß in eqn. 2.5 is linked with a series of quantities:

$$ß = 2\pi/\lambda_\varepsilon = 2\pi\sqrt{\varepsilon_r}/\lambda_0 = 2\pi f\sqrt{\varepsilon_r}/c = \omega/\upsilon_\varepsilon \qquad (2.5)$$

λ_0 = wavelength in free space
λ_ε = wavelength in the dielectric
c = speed of light ($3 \cdot 10^8$ m/s)

υ_ε = wave velocity in the dielectric
f = frequency
ε_r = relative dielectric constants

The transmission constant is derived according to eqns. 2.1 and 2.2 as $\gamma = \alpha + j\beta$. For each transmission line of electromagnetic waves, the quotient of voltage and current waves is a characteristic quantity. The quotients of 'departing' voltage and current waves are designated as line wave impedance Z_L:

$$Z_L = U_{fwd}/J_{fwd} \qquad (2.6)$$

The wave impedance is a constant value, irrespective of the line length z. Equations 2.1, 2.2, 2.4 – 2.6 provide the links between the three constants α, β, and Z_L. A transmission line is characterised by these three quantities.

The task of transmission lines is to transfer electrical energy and electrical signals. In the process it should be guaranteed that electrical voltages and currents at given frequencies are transferred as far as possible with low attenuation, and free of interference and distortion. The problems of power transfer, in other words what maximum power can be transferred, can be identified from the relationship shown in Figure 2.5. The electric field E between the upper and lower line must not exceed a maximum value (in dry air $E_{max} = 30$ kV/cm).

$$\underline{U}(z=0) \cdot \underline{J}(z=0) = U_{fwd} \cdot J_{fwd} = U^2_{fwd}/Z_L = P_{fwd} \qquad (2.7)$$

With a given type of line, the value of Z_L is determined (usually with coaxial cables and strip lines, values are about $Z_L = 50\text{–}75\ \Omega$). The upper power limit depends on the maximum permissible voltage. If the voltage is increased more, one will have a voltage sparkover; therefore, a limit is given for the maximum power (P_{max}). This power limit is of extreme importance in general in cases of pulse transfer, such as in radar technology.

The line behaves differently when transferring sustained power. The limit of the maximum sustained power transfer ($\bar{P}_{max}$) is determined by the Joule heat losses, in other words by the size of the attenuation constant α. Only a limited amount of Joule heat can be dispersed per unit length of line. At high power absorption this will lead to high temperatures in the line structures. Depending on the materials used, temperatures of more than 100°C are in practice only permissible in special cases.

A distortion-free signal transfer is another important criterion for a line. Electrical signals are characterised by their spectral energy distribution and their bandwidth. The wave components can be described by eqns. 2.1 and 2.2, or by the time-dependent voltage and current values u (t, z) and i (t, z), $(u(t,z) = \text{Re}[\sqrt{2}\underline{U}\ (z)\ e^{j\omega t}])$:

$$u(t,z) = \hat{U}_{fwd} \cos(\omega t - \beta z) \text{ with } \hat{U}_{fwd} = U_{fwd}\sqrt{2};\ \beta = 2\pi/\lambda_\varepsilon \qquad (2.8)$$

$$i(t,z) = \hat{J}_{fwd} \cos(\omega t - \beta z) \text{ with } \hat{J}_{fwd} = J_{fwd} \sqrt{2} \qquad (2.9)$$

In these equations the phase dependency of the voltages and currents has been left out of consideration. The phase could be included in the evaluation by the phase angle φ in the round brackets $(\omega t - \beta z + \varphi)$ or by $\varepsilon^{j\varphi}$ in eqns. 2.1 and 2.2.

A signal consists of a considerable number of voltages and currents $u(t,z)$ and $i(t,z)$ at various different frequencies. A necessary condition for distortion-free transmission is the frequency-independent propagation of the individual waves.

Figure 2.6 represents the process of an attenuated voltage wave according to eqn. 2.8. In this context, $T = 1/f$ is the period of oscillation, and υ_ε is the speed of propagation. The speed of propagation υ_ε should be of equal value for all the individual waves occurring, throughout the entire frequency band. A frequency transformation $d\omega$ should, therefore, not cause any change in the speed $\Delta\upsilon_\varepsilon$.

$$d\upsilon_\varepsilon/d\omega = 0 \qquad (2.10)$$

It follows from this, with eqn. 2.5, that

$$d\left(\frac{\omega}{\beta}\right)/d\omega = (1 - \frac{\omega}{\beta}\frac{d\beta}{d\omega})/\beta = (1 - \upsilon_\varepsilon \frac{d\beta}{d\omega})/\beta = 0 \qquad (2.11)$$

With $\beta = \omega\sqrt{\varepsilon_r}/C$, for a distortion-free transmission:

$$\frac{d\beta}{d\omega} = \frac{\sqrt{\varepsilon_r}}{C} + \frac{\omega}{C}\frac{d\sqrt{\varepsilon_r}}{d\omega} = \frac{1}{\upsilon_\varepsilon} \text{ at } \frac{d\sqrt{\varepsilon_r}}{d\omega} = \frac{1}{2\sqrt{\varepsilon_r}}\frac{d\varepsilon_r}{d\omega} = 0 \qquad (2.12)$$

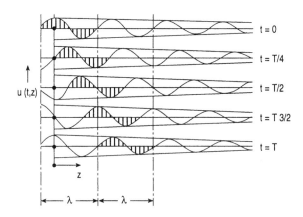

Figure 2.6   *Propagation process of attenuated voltage wave proceeding in z-direction.*

Equation 2.12 is only fulfilled if ε_r is frequency independent. This condition is fulfilled with a good degree of approximation with virtually all Lecher wires.

Problem 1: A power $P = 50$ W is fed into an adapted reflection-free coaxial cable at 1000 MHz. The values are $D = 3.01$ mm; $d = 0.9$ mm; $\varepsilon_r = 2.1$; $\alpha = 0.7$ dB/m. Cable length $l = 10$ m.

    1.1   What is the wave impedance of the cable?

    1.2   What power loss V is incurred along the cable length in the form of heat given off to the environment?

    1.3   What time-dependent voltage is derived at the end of the transmission line?

Solution: 1.1 $Z_L = 50\ \Omega$ (Figure 2.1)

    1.2   Power at the end of the line:

$$P(l) = (U_{fwd}\, e^{-\alpha l})2/Z_L = Pe - 2\alpha l = 10\ W\ \ V = 40\ W$$

    1.3 $\hat{U}_{fwd} = \sqrt{2} \cdot U_{fwd} = \sqrt{2}\ \sqrt{PZ_L} = 70.7$ V

           $\lambda = \lambda_\varepsilon = 0.3/\sqrt{2.1}$ m $= 0.21$m; $e - \alpha l = 0.45$

           $u(t,z) = 31.6\cos(\omega t\ \ 110°)$ V

2.2.2 *Lines with reflections*

As an extension to the reflection-free case considered in Section 2.2.1, now examine wave reflections caused by mismatching at the end of the line. If the ratio of voltage wave to current wave of the forward-moving wave, at the end of the line is interfered with by the fact that $U_{fwd}/J_{fwd} = Z_L$ is not identical to the matching resistance R_a, then we have a wave reflection. At $z = l$, we have a reflected wave, i.e. one which is travelling in the negative z-direction. The size of this wave, which has its origin at $z = l$, and which propagates contrary to the forward wave (eqn. 2.1), is derived as

$$U_{refl}(z) = U_{refl}\, e^{\gamma z} \qquad\qquad (2.13)$$

The time-dependent value is

$$U_{refl}(t,z) = \hat{U}_{refl}\, e^{\alpha z} \cos(\omega t + \beta z + \varphi). \qquad\qquad (2.14)$$

In accordance with the line length l, a phase angle φ is associated with the reflected wave. A so-called reflection factor is defined at the end of the line:

$$\underline{r} = U_{ref(l)}/U_{fwd(l)} \qquad\qquad (2.15)$$

This value of $\underline{r}$ is related to the terminating impedance $\underline{Z}_a$:

$$\underline{\mathfrak{r}} = \frac{\underline{Z}_a - Z_L}{\underline{Z}_a + Z_L} \tag{2.16}$$

If the terminating impedance $\underline{Z}_a$ is a real value R_a, then a real value is likewise derived for $\mathfrak{r}$:

$$\mathfrak{r} = \frac{R_a - Z_L}{R_a + Z_L} \tag{2.17}$$

With the lines in a steady-state condition, in other words, with no extremely short voltage pulses, every complex impedance $\underline{Z}_a$ can be transformed into a resistance R_a and a line section (Z_L) of length l_o (Figure 2.7):

$$\underline{Z}_a = Z_L(R_a/Z_L + j\tan\beta l_o)/(1 + j(R_a/Z_L)\tan\beta l_o) \tag{2.18}$$

Instead of calculating this expression, in practice it is sufficiently precise to use the Smith chart. The complex resistance $\underline{Z}_a$ is to be determined for $R_a = 20\,\Omega$, on a $50\,\Omega$ cable. The related line length l_o/λ_ϵ is 0.1.

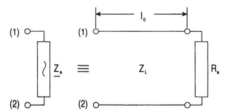

Figure 2.7   *Representation of complex resistance as a transmission line with real terminating resistance*

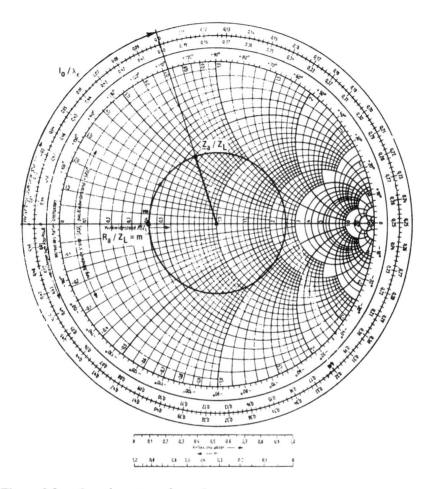

Figure 2.8 *Impedance transformation*

Figure 2.8 shows the transformation path. The line length l_0/λ is transformed in the Smith diagram to the right in the direction of the generator, onto an arc of a circle. In this context, an arc sector from 180° to 108° is selected. The arc intersects the two limbs of the sector at the point R_a/Z_L on the real axis, and at Z_a/Z_L in the diagram. The midpoint of the circle is the matching point $R/Z_L = 1$. The radius of the circle corresponds to the value of the reflection factor $|\underline{r}|$. The value $Z_a/Z_L = 0.57 + j0.57$ is read off from the diagram, and the value derived for the complex terminating resistance is $Z_a = (28.5 + j\,28.5)\Omega$.

Formulated another way, and returning to the initial problem, the complex resistance Z_a can therefore be described according to Figure 2.7 by connecting a line with the wave impedance Z_L, the line length l_0, and with the resistance R_a. For a total length l of any desired line, with a random complex resistance, it is assumed hereafter that a real termination value R_a obtains. The extension of the line by the length l_o in this context plays only a subordinate role.

The quotient R_a/Z_L is frequently designated as the ripple m, and the reciprocal value as a standing wave ratio (SWR) s:

$$m = 1/s = R_a/Z_L = U_{min}/U_{max} \qquad (2.19)$$

The ripple m is also related to the values U_{min} and U_{max} of Figure 2.9 along the line.

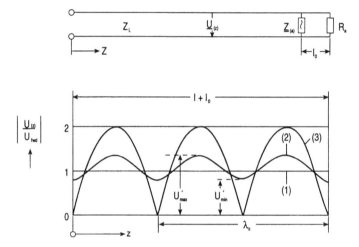

Figure 2.9   *Voltage distribution: wave reflection at various resistances*

    (1) $R_â = Z_L$
    (2) $R_â = Z_L/2$
    (3) $R_â = 0$

In this context, it is a precondition that the value R_a is smaller than R_L. For example, let the terminating resistance $Z_a = R$ be a real value greater than Z_L. By means of a transformation about $l_o/\lambda = 0.25$, a value is derived $R_a/Z_L = Z_L/R$. This is known as a $\lambda/4$ transformation. In Figure 2.10, the transformation is carried out for $R = 100\ \Omega$, $Z_L = 50\ \Omega$, and $l_o = \lambda/4$. In this context, $R_a = 25\ \Omega$.

Ripple m and reflection factor r inter-relate via eqns. 2.17 and 2.19. Both values are real values. For the reflection factor r:

$$r = \frac{m - 1}{m + 1} \qquad (2.20)$$

From the assumption $R_a < Z_L$ it follows that r is a negative real value.

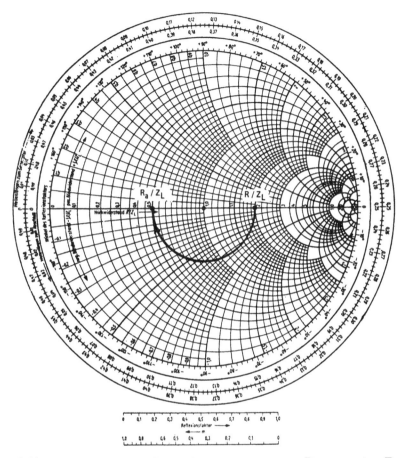

Figure 2.10 *Quarter wavelength transformation: resistance R greater than Z_L is transformed into resistance R_a smaller than Z_L*

In practice, what is of interest to most people is only the value of the reflection factor. The input resistance Z_a of a line with the wave impedance Z_L and the variable line length l_0 in the Smith diagram lies on an arc. In this situation, the line is terminated with a constant real resistance value R_a. The midpoint of the arc is the midpoint (matching point 1) of the Smith diagram. For all complex impedances Z_a which lie on the arc, the relationship: $r = (R_a-Z_L)/(R_a+Z_L)$ applies irrespective of the value l_0/λ.

In other words, in a Smith diagram one can write the constant parameters of the reflection factor r onto the circular locus of Z_a. We then come to the statement that all the values of Z_a with a fixed reflection factor are interconnected. On the other hand, according to eqn. 2.20, the reflection factor r is linked to the ripple m. This means that, instead of the parameter r, the parameter of the ripple m can also be written onto the circular locus of Z_a. Concentric circles around the matching point are called m-circles. Thus, for example, the locus Z_a/Z_A for $R_a/Z_L = 0.5$ is an m-circle with the value $m = 0.5$, or an r-circle with the value $r = (1-m)/1+m) = 0.33$. According to eqn. 2.16, the complex reflection factor r depends on the complex resistance value Z_a/Z_L. The value of r can be determined in the diagram; for example, in Figure 2.8, this value is $r = 0.43\ e^{j\ 108°}$.

For the wave components travelling backwards and forwards, from eqns. 2.1 and 2.13,

$$U(z) = U_{fwd}(z) + U_{ref}(z);\ \text{with } r = U_{refl}(l)/U_{fwd}(l) = J_{refl}(l)/J_{fwd}(l) \qquad (2.21)$$

$$J(z) = J_{hin}(z) - J_{refl}(z)$$

If we calculate the outgoing and reflected voltage waves for $z = l' = l + l_o$, with the aid of eqns. 2.1, 2.13 and 2.20, and insert the result in eqn. 2.21, then

$$U(z) = U_{fwd}\ (e^{\gamma(l'-z)} + r\ e^{-\gamma(l'-z)}) \qquad (2.22)$$

In the same way, one can derive the current $J(z)$:

$$J(z) = J_{fwd}\ (e^{\gamma(l'-z)} - r\ e^{-\gamma(l'-z)}) \qquad (2.23)$$

In the case of high-frequency lines, one can assume with a good degree of approximation, that there will be no loss on short lines. In this case, $\alpha = 0$ and $\gamma = j\beta$. The value of the voltage $U(z)$ from eqn. 2.22, related to U_{fwd} then gives

$$|U(z)|/U_{fwd} = \sqrt{(1+r^2) + 2r\ \cos\ ((4\pi/\lambda)(l'-z))} \qquad (2.24)$$

With the help of this equation, the voltage characteristic on a line can be determined. In Figure 2.9 the voltage distributions along the line are shown for various reflection factors. In the case of total reflection $(r = -1)$, standing waves are derived with zero settings and maxima $(m = U_{min}/U_{max} = 0)$. With reflection-free transmission $(r = 0)$ the voltage at the different points of the line is constant $(m = U_{min}/U_{max} = 1)$.

Example 2:   A generator $(R_i = 60\ \Omega,\ U_o = 10\ V)$ feeds an antenna, at 200 MHz via a no-loss line $(Z_L = 60\ \Omega;\ \varepsilon_r = 2.25)$. The antenna has a complex input resistance $Z_a = (120 - j\ 90)\ \Omega$. The following are sought:

2.1   The power P_{fwd} fed into the line
2.2   The equivalent values R_a, l_o for the antenna
2.3   Ripple m, standing wave ratio s, reflection factor r
2.4   Reflected power P_{refl}
2.5   Effective value of the voltage minimum and the voltage maximum along the line
2.6   Complex reflection factor at the antenna input

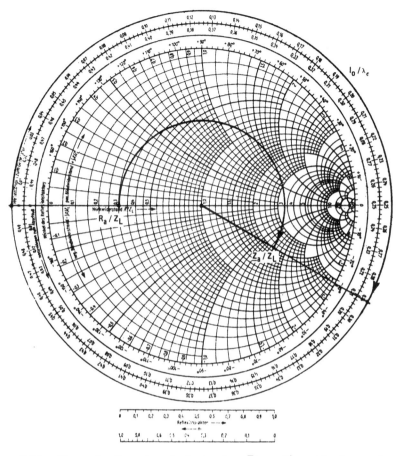

Figure 2.11 *Determination of equivalent values* R_a *and* I_0, *with given values of* $\underline{Z}_a$ *and* Z_L

Solution:

2.1 $P_{fwd} = U_{fwd}^2/Z_L = U_0^2/4; R_i = 0.42$ W

2.2 $\underline{Z}_a/Z_L = 2 - j1.5$; see eqn. 2.11; $R_a = 18$ Ω; $\lambda_\varepsilon = 1$ m; $I_0 = 0.291$ m

2.3 $m = 0.3$; $s = 3.3$; $r = 0.54$

2.4 $P_{refl} = U_{refl}^2/Z_L = r^2 \cdot P_{fwd} = 0.12$ W

2.5 $U_{min} = U_{fwd} - U_{refl} = U_{fwd}(1-r) = 2.31$ V
$U_{max} = U_{fwd} + U_{refl} = U_{fwd}(1+r) = 7.69$ V
$m = U_{min}/U_{max} = 0.3$

2.6 $\underline{r} = (\underline{Z}_a/Z_L - 1)/(\underline{Z}_a/Z_L + 1) = re^{-j\,330.5°} = 0.54e^{-j\,330.5°}$

2.3 Line transformation

The quotient of $\underline{U}(z)$ and $\underline{J}(z)$ from eqns. 2.22 and 2.23 is the complex line resistance $\underline{Z}(z)$. With a no-loss line we derive

$$\underline{Z}(z) = \underline{U}(z)/\underline{J}(z) = Z_L\,(1 + re^{-j2\beta(l'-z)})/(1 - re^{-2\beta(l'-z)}) \qquad (2.25)$$

In this situation, for r we are to use the real value according to eqn. 2.17. This equation can be written differently by transformation and by substituting m for r (eqn. 2.20)

$$\underline{Z}(z) = Z_L\,(m + j\,\tan\beta(l'-z))/(1 + jm\,\tan\beta(l'-z)) \qquad (2.26)$$

2.3.1 Series connection of lines

With high-frequency lines, series connections of lines with different wave impedances are frequently created. Figure 2.12 shows a series connection of lines. An equivalent connection of a two-core line is taken as the basis, and we usually describe all possible types of lines by means of this equivalent circuit. In the example in question, these are co-axial cables, striplines, and hollow conductors.

The intention in the following example is to determine the complex input impedance $\underline{Z}_{in}$ of the series connection of lines according to Figure 2.12. The complex resistance $\underline{Z}_a$ is initially related to the line Z_{L1} by

$$\underline{Z}'_a = \frac{\underline{Z}_a}{Z_{L1}}.$$

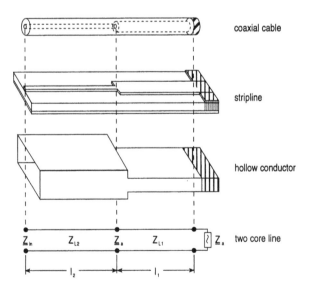

Figure 2.12   *Series connection of lines with different wave impedances*

The new complex value $\mathbf{Z'_b}$, related to Z_{L1}, is the value of $\mathbf{Z'_a}$ transformed over the line length l_1 (see Figure 2.13).

By denormalising, we derive $\mathbf{Z_b} = \mathbf{Z'_b} \cdot Z_{L1}$. As the next step, a transformation is carried out over the line length l_2. The value of $\mathbf{Z_b}$ is now related to Z_{L2}. In Figure 2.13 we obtain the new value of $\mathbf{Z'_{bnew}}$. $\mathbf{Z'_{in}}$ is derived by the transformation l_2/λ in the direction of the source.

A further denormalisation leads to the value of $\mathbf{Z_{in}}$.

Example: The aim is to determine the input impedance of a waveguide circuit. It is intended that an impedance $\mathbf{Z_a} = (100 + j\,100)\ \Omega$ is to be transformed over line 1, (length $l_1/\lambda = 0.1$, principal wave impedance $Z_{L1} = 50\ \Omega$) and over line 2 (length $l_2/\lambda = 0.2$, wave impedance $Z_{L2} = 100\ \Omega$) according to Figure 2.13.

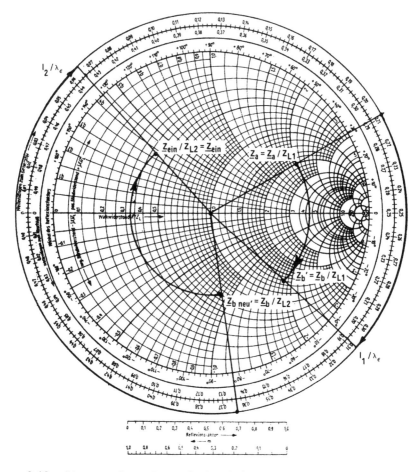

Figure 2.13   *Line transformation in the Smith diagram with series connections of lines*

With this example, the type of waveguide is not significant. In this situation, it could be a coaxial cable, a stripline, a hollow conductor, or a two-core line; the result will in all cases be basically the same. The input resistance Z_{in} of the circuit can be determined according to the transformation path of Figure 2.13. Specifically, we derive:

$$Z_a/Z_{L1} = 2 + 2j; \; Z_b/Z_{L1} = 1.3 - j\,1.8; \; Z_b = (65 - j\,90)\;\Omega;$$

$$Z_b/Z_{L2} = 0.65 - j\,0.9$$

$$Z_{in}/Z_{L2} = 0.38 + j\,0.38; \; Z_{in} = (38 + j\,38)\;\Omega$$

2.3.2 Parallel connection of lines

Interface circuits are frequently created by means of parallel connections. Figure 2.14 shows a number of such interface circuits. A basic criterion in this context is whether such parallel-connected lines are connected with a short-circuit ($R_a = 0$) or in no-load operation ($R_a \rightarrow \infty$). Lines such as these are referred to as spur lines.

Figure 2.14 shows the case of $R_a = 0$. In the case of interface circuits of microstriplines, however, a no-load spur line is preferred in practice because of the simpler structure involved.

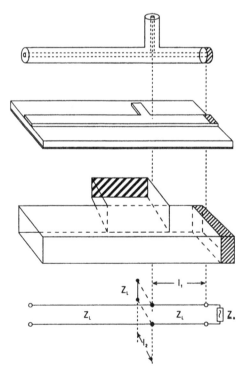

Figure 2.14 *Lines connected in parallel*

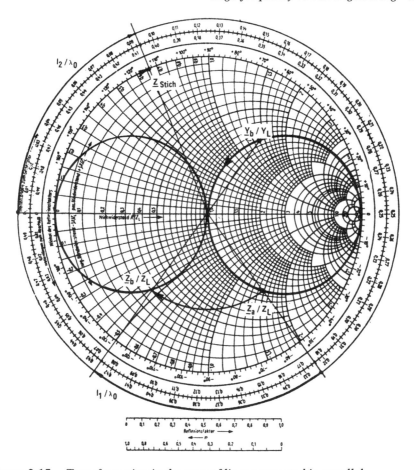

Figure 2.15 *Transformation in the case of lines connected in parallel*

Example:　The intention is to create an interface circuit by means of a no-load spur line (length l_2) at the location l_1, according to Figure 2.14 for $Z_a = (50 - j\,75)\ \Omega$. We are given: $Z_L = 50\ \Omega$, $\lambda_o = 20$ cm. How large are l_1 and l_2?

Solution:　$\underline{Z}_a$ is related to $\underline{Z}_L$: $\underline{Z}'_a = 1 - j\,1.5$. According to Figure 2.15, the transformation is carried out to the interface circuit via $l_1/\lambda_o = 0.102$. In this context, we obtain $\underline{Z}'_b = 0.3 - j\,0.47$. In view of the fact that a parallel circuit of lines can be best obtained in the Smith diagram by an addition of admittance values, we determine from $\underline{Z}'_b$ the related admittance value $\underline{Y}'_b = 1/\underline{Z}'_b$. In this situation, we derive $\underline{Y}'_b = 1 + j\,1.5$.

The aim of the adaptation measure is the adaptation point $\underline{Y}_{ges} = 1$. By the addition of $\underline{Y}_{st} = 1/\underline{Z}'_{st} = -j\,1.5$ ($\underline{Z}'_{st} = j\,0.666$) we derive $\underline{Y}'_{ges} = \underline{Y}'_b + \underline{Y}'_{st} = 1$.

The line length $l_2 = 8.26$ cm is obtained according to Figure 2.15 by transformation of the no-load resistance ($R_a = \infty$) by the value $l_2/\lambda_0 = 0.34$.

2.4 Connection between scattering parameters and line sizes

Linear high-frequency components are in many cases described by scattering parameters. In Figure 2.16, a high-frequency component of this type is shown connected between two lines with the same wave impedance Z_L. In the line section on the primary side (1), the values derived by the wave equations (eqn. 2.21) at the point $z = l_1$, in other words at the input of the quadripole, are $U_{fwd}(l_1)$ and $U_{ref}(l_1)$. We designate these voltage values as $U_{fwd}(l_1) = U_{1fwd}$ and $U_{ref}(l_1) = U_{1ref}$.

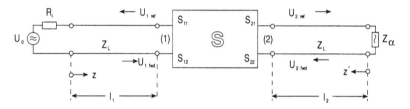

Figure 2.16   *Outgoing and reflected voltage waves at input and output of quadripole*

Likewise one can designate the output waves of the quadripole with U_{2fwd} and U_{2ref}. The S-parameter describes the relationship of the voltage waves before and after the quadripole.

For the special case in which lines are assumed on the primary and secondary side with the same wave resistances Z_L, we derive a simple possibility for interconnecting the waves:

$$U_{1ref} = S_{11}U_{1fwd} + S_{12}U_{2fwd} \tag{2.27}$$

$$U_{2ref} = S_{21}U_{1fwd} + S_{22}U_{2fwd}$$

The intention is to determine the scattering parameters for a simple symmetrical quadripole according to Figure 2.17 with the complex resistances Z. The line on the secondary side (length l_2) is terminated free of reflection ($U_{2fwd} = 0$) with $R_a = Z_L$. From eqn. 2.27 we accordingly derive

$$U_{1ref} = S_{11}U_{1fwd} \qquad\qquad U_{2ref} = S_{21}U_{1fwd}. \tag{2.28}$$

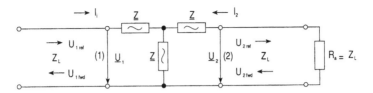

Figure 2.17 *S-parameter calculation with quadripole in T-circuit*

The reflection factor $\underline{r}$ at the quadripole input is related with the scattering parameter $\underline{S}_{11}$ according to eqns. 2.15, 2.16 and 2.28. An input impedance $\underline{Z}_{in}$ is derived for the example:

$$\underline{Z}_{in} = \frac{3\underline{Z}^2 + 2\underline{Z} \cdot Z_L}{2\underline{Z} + Z_L} . \tag{2.29}$$

In this context we derive for $\underline{r}$ and $\underline{S}_{11}$:

$$\underline{r} = \underline{S}_{11} = \frac{\underline{Z}_{in} - Z_L}{\underline{Z}_{in} + Z_L} = \frac{\underline{U}_{1ref}}{\underline{U}_{1fwd}} = \frac{3\underline{Z}^2 - Z_L{}^2}{3\underline{Z}^2 + 4\underline{Z} \cdot Z_L + Z_L{}^2} \tag{2.30}$$

In the calculation of $\underline{S}_{21}$, proceed in such a way that the voltage $\underline{U}_1$ is made up from $\underline{U}_{in}$ and $\underline{U}_{1ref}$ according to eqn. 2.21. Because $\underline{U}_{in} = 0$, the voltage $\underline{U}_2$ is identical to $\underline{U}_{2ref}$.

$$\underline{U}_1 = \underline{U}_{1fwd} + \underline{U}_{1ref} = \underline{U}_{1fwd}(1 + \underline{S}_{11}); \quad \underline{U}_2 = \underline{U}_{2ref}. \tag{2.31}$$

From Figure 2.17, the voltage ratio is derived via the mesh equation:

$$\underline{U}_2/\underline{U}_1 = \underline{U}_{2ref}/\underline{U}_{1fwd}(1 + \underline{S}_{11}) = \underline{S}_{21}/(1 + \underline{S}_{11}). \tag{2.32}$$

This resolves to

$$\underline{S}_{21} = \frac{2\underline{Z} \cdot Z_L}{4\underline{Z} \cdot Z_L + 3\underline{Z}^2 + Z_L{}^2} . \tag{2.33}$$

Due to the symmetry of the quadripole, the reversibility of the circuit is a possibility. From this follows $\underline{S}_{22} = \underline{S}_{11}$; $\underline{S}_{12} = \underline{S}_{21}$.

The advantage of introducing S-parameters is particularly clear if we are considering hollow conductor quadripoles (Figure 2.18). In the case of hollow conductors, voltages and currents are values that are not directly measurable. We introduce here the outgoing and reflected wave voltages discussed previously.

By contrast to the situation with Lecher lines, in the case of hollow conductors only what are known as fictional voltages can be defined. It is, however, possible to transfer from hollow conductor waves immediately to voltage waves (eqns. 2.21, 2.22), if we consider only the wave parameters in the description of lines and components. Figure 2.19 shows how the S-parameters of any random hollow conductor component (HB) can be determined.

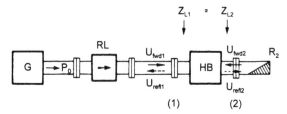

Figure 2.18 *Hollow conductor quadripole (two-port network) (HB) in the wave field of outgoing and reflected voltages*

RL = isolator; G = generator

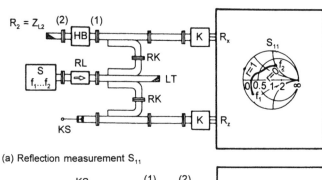

(a) Reflection measurement S_{11}

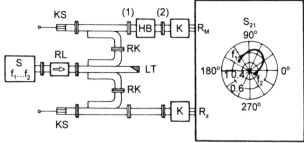

(b) Transmission coefficient measurement S_{21}

Figure 2.19 *Measuring S-parameters with network analysis device*

S = sweep signal transmitter; RL = isolator; RK = directional coupler;
LT = load impedance (reflection-free); HB = hollow conductor component;
KS = short-circuit line; $R_2 = Z_{L2}$ = reflection-free terminating resistance;
K = frequency converter; R_x = reflected voltage; R_z = reference voltage;
R_M = measuring circuit voltage

The basic principle of the method is the comparison measurement of voltage waves. A wave which comes from a component, in other words from a defined input or output plane of a quadripole, is compared with a second wave. This second wave, in this context, is identical with regard to phase and amplitude to a wave which comes from the input, for example, of a totally reflected quadripole. If the output of the quadripole to be measured is terminated free of reflection, the reflection factor of the input is, according to the definition, identical to S_{11}.

An evaluation circuit (network analyser) compares the reference voltage with the voltage from the quadripole. The phase and amplitude difference provides a statement about the value and phase of $\underline{S}_{11}$ (or S_{22} in the case of HB being reversed). Similarly, according to Figure 2.19, the voltage transformation factor $\underline{S}_{21}$ is measured (or $\underline{S}_{12}$ if the component is reversed).

Example 3: A quadripole according to Figure 2.17 is supplied from a generator with a power of $P_{fwd} = 600$ mW. Given are: $Z_L = 60\ \Omega$ no-loss line, $R_i = R_a = 60\ \Omega$. $\underline{Z} = 60\ \Omega$.

We are seeking:

3.1 $\underline{S}_{11}, \underline{S}_{22}, \underline{S}_{12}, \underline{S}_{21}$
3.2 $\underline{U}_{1fwd}, \underline{U}_{1ref}, \underline{U}_{2fwd}, \underline{U}_{2ref}$
3.3 Input reflection of the quadripole $\underline{r}$
3.4 Power consumption of R_a
3.5 Reflected power P_{ref}
3.6 Power loss P_v in the quadripole
3.7 Value of the voltage characteristic on the line before the quadripole

Solution: 3.1 $\underline{S}_{11} = \underline{S}_{22} = 1/4$; $\underline{S}_{12} = \underline{S}_{21} = 1/4$ according to eqns. 2.30 and 2.33

3.2 $\underline{U}_{1fwd} = \sqrt{P_{fwd}\ Z_L} = 6$ V; $\underline{r} = \underline{S}_{11} = 1/4$

3.3 $\underline{U}_{1ref} = \underline{r}\ \underline{U}_{1fwd} = 1.5$ V
$\underline{U}_{2fwd} = 0$; $\underline{U}_{2ref} = \underline{S}_{21} \cdot \underline{U}_{1fwd} = 1.5$ V

3.4 Power consumption $R_a : P_a = \underline{U}_{2ref}^2/R_a = 37.5$ mW
3.5 Reflected power $P_{ref} = r^2\ P_{fwd} = 37.5$ mW

3.6 Quadripole power loss $P_v = P_{fwd} - P_{ref} - P_a = 525$ mW
3.7 Voltage characteristic according to eqn. 2.24:

$$|\underline{U}(z)| = U_{fwd}\ \sqrt{(1 + r^2) + 2r\ \cos\ (4\pi\ (l_1 - z)/\lambda)}$$
$$= 6\ V\ \sqrt{17/16 + 0.5\ \cos\ (4\pi\ (l_1 - z)/\lambda)}$$

2.5 Coaxial cable

The coaxial cable is the type of connection cable most frequently used for high-frequency transmissions. Its areas of application are mainly as follows.

2.5.1 Wide-area transmission technology

The state of the transmission art at present is characterised predominantly by coaxial cable. The majority of wide-area connections run over coaxial cables (Z_L = 75 Ω) (Figure 2.20). The carrier frequency systems, as shown in Figure 2.21, are in frequency ranges between 60 kHz and 60 MHz. Thus, for example, coaxial conductors are used for the analogue system V 2700, the inner conductors of which have a diameter of 2.6 mm and the outer conductors a diameter of 9.5 mm. With these cables, a relatively low attenuation of 0.08 dB/100 m is achieved at 12.4 MHz. The interim space between the inner and outer conductors is largely free of dielectric materials. Only thin discs are fitted as spacers. The wave impedance Z_L is 75 Ω. For the mean effective dielectric constant, the value of ε_r = 1.07 is derived in this context (Figure 2.20). In the wide-area sector, the somewhat thinner 1.2/4.4 mm coaxial cable is also often used, with attenuation values of about 0.16 dB/100 m at 12.4 MHz.

Figure 2.20   *Six-stage coaxial cable, Cu 2.6/9.6 mm*

In addition to analogue systems, digital transmission systems are being increasingly used. A frequency diagram is shown in Figure 2.22. The digital signals are as a rule transferred over the same coaxial cable lines as are used for analogue signals.

2.5.2 Antenna cables

Coaxial cables are used predominantly as the feed lines for television and radio reception antennas. Figure 2.23 shows screened antenna cables. These cables are used in the reception ranges L M S U (long, medium, short, and ultra-short wave), and in the ranges F I to F V (television channels I to V). The attenuations in dB are given in lengths of 100 m.

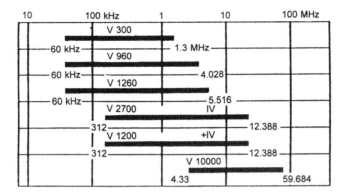

Figure 2.21   *Frequency distribution of the most important analogue transmission systems*

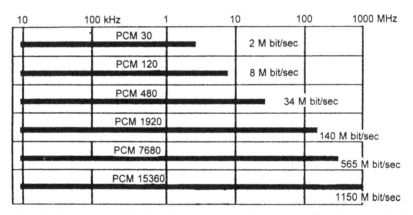

Figure 2.22   *Frequency distribution of the most important digital transmission systems*

2.5.3 High-frequency cables

High-frequency cables are used, depending on the frequency range, as the connection lines for equipment, apparatus, and circuits. A variety of demands are placed on the cable, such as flexibility, good attenuation properties, robustness, weatherproof properties, dielectric strength, and transmission capacity for high sustained outputs.

Table 2.1 lists a series of important HF cables. The HF operating voltage given in the table relates to the highest peak voltage permitted in the channel (pulse voltage) U_{sp}. For a maximum electrical disruptive field strength E_d, the peak voltage derived is

$$U_{sp} = E_d \cdot \frac{d}{2} \frac{Z_L \cdot \sqrt{\varepsilon_r}}{60} \tag{2.34}$$

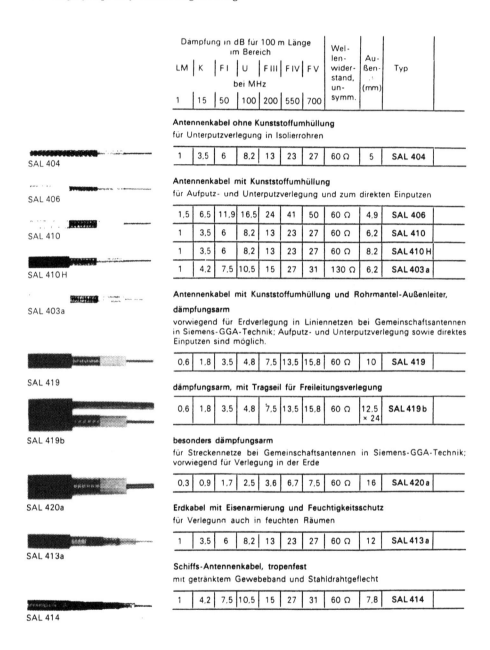

Dämpfung in dB für 100 m Länge im Bereich							Wellen-widerstand, unsymm.	Außen- (mm)	Typ
LM	K	F I	U	F III	F IV	F V			
bei MHz									
1	15	50	100	200	550	700			

Antennenkabel ohne Kunststoffumhüllung
für Unterputzverlegung in Isolierrohren

1	3,5	6	8,2	13	23	27	60 Ω	5	SAL 404

Antennenkabel mit Kunststoffumhüllung
für Aufputz- und Unterputzverlegung und zum direkten Einputzen

1,5	6,5	11,9	16,5	24	41	50	60 Ω	4,9	SAL 406
1	3,5	6	8,2	13	23	27	60 Ω	6,2	SAL 410
1	3,5	6	8,2	13	23	27	60 Ω	8,2	SAL 410 H
1	4,2	7,5	10,5	15	27	31	130 Ω	6,2	SAL 403 a

Antennenkabel mit Kunststoffumhüllung und Rohrmantel-Außenleiter,

dämpfungsarm

vorwiegend für Erdverlegung in Liniennetzen bei Gemeinschaftsantennen in Siemens-GGA-Technik; Aufputz- und Unterputzverlegung sowie direktes Einputzen sind möglich.

0,6	1,8	3,5	4,8	7,5	13,5	15,8	60 Ω	10	SAL 419

dämpfungsarm, mit Tragseil für Freileitungsverlegung

0,6	1,8	3,5	4,8	7,5	13,5	15,8	60 Ω	12,5 × 24	SAL 419 b

besonders dämpfungsarm

für Streckennetze bei Gemeinschaftsantennen in Siemens-GGA-Technik; vorwiegend für Verlegung in der Erde

0,3	0,9	1,7	2,5	3,6	6,7	7,5	60 Ω	16	SAL 420 a

Erdkabel mit Eisenarmierung und Feuchtigkeitsschutz
für Verlegung auch in feuchten Räumen

1	3,5	6	8,2	13	23	27	60 Ω	12	SAL 413 a

Schiffs-Antennenkabel, tropenfest
mit getränktem Gewebeband und Stahldrahtgeflecht

1	4,2	7,5	10,5	15	27	31	60 Ω	7,8	SAL 414

SAL 404
SAL 406
SAL 410
SAL 410 H
SAL 403a
SAL 419
SAL 419b
SAL 420a
SAL 413a
SAL 414

Figure 2.23   *Antenna cables*

Table 2.1 *High-frequency cables*

HF cables

Type	Impedance Ω	HF operating voltage V	Attenuation in dB/100 m at 100 MHz	Capacitance pf/m	External dia. mm	Notes
RG 22	95	–	11.50	52.0	10.30	twin cable
RG 58 C	50	1900	16.07	98.4	4.95	
RG 59 B	75	2300	11.15	68.9	6.15	
RG 62 A	93	750	8.86	44.3	6.15	
RG 108 A	78	1000	24.60	77.1	5.97	twin cable
RG 174	50	1500	29.19	88.4	2.54	subminiature
RG 213	50	5000	6.23	96.8	10.29	
RG 214	50	5000	7.54	96.8	10.80	
RG 218	50	11000	3.11	96.8	22.10	
RG 223	50	1900	15.74	96.8	5.49	
21-204	50	1900	–	96.8	7.24	triaxial cable
21-541	73	2300	16.40	70.5	6.14	low-noise
21-597	75	2500	19.02	65.6	3.81	subminiature
21-738	52	2000	–	96.8	4.95	triaxial cable
AL 0.8/3.2 L	60	3000	14.60	84.0	4.90	twisted leads
HFE 1.5/6.5	60	1500	6.60	84.0	8.80	bright Cu wire
HFE 1.5/6.5 L	60	6000	7.70	84.0	8.80	leads
HFE2.3/10	60	9500	4.60	84.0	12.60	bright Cu wire

Teflon HF cables

RG 178 B	50	1000	45.92	93.0	1.90	
RG 187 A	75	1200	31.81	63.0	2.79	
RG 188 A	50	1200	37.39	95.0	2.79	
RG 196	50	1000	45.92	93.0	2.03	
RG 196 A	95	1500	45.92	49.0	3.94	

The maximum field strengths in a coaxial cable occur in the inner conductor. One can reckon on maximum permissible values of about $E_d = 30\ kV/cm$.

Figure 2.24 shows a compilation of data relating to various different cable types. Here, the attenuation values are shown in dB per 100 m. In addition, the curves of the HF continuous output are shown as a function of frequency.

2.6 Hollow conductors

For frequencies above about 20 GHz, it is recommended that hollow conductors be used as the transmission lines. It is also recommended that hollow conductors be used for high power ratings (above 100 W), in the frequency range above about 3 GHz. Hollow conductors are also to be preferred to other types of line if, at high frequencies, considerable value is being set on extremely low attenuation. Table 2.2 shows some of the most important square hollow conductors.

According to Figure 2.25, a basic trend can be seen with regard to the attenuation values to be anticipated in the different transfer ranges of hollow conductors. A constant increase in mean attenuation can be seen across the entire frequency range up to 150 GHz.

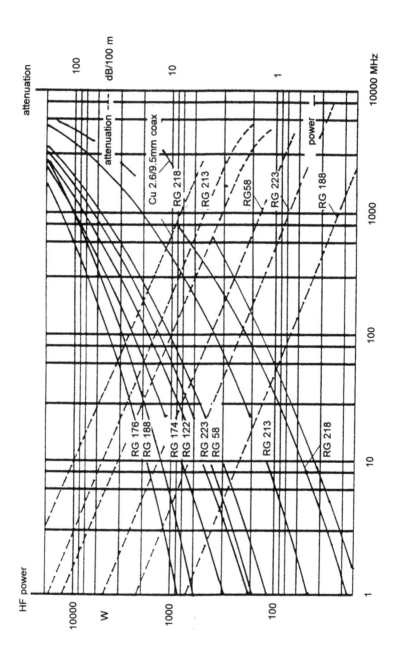

Figure 2.24 *Frequency dependency of attenuation and maximum permissible HF continuous output*

Table 2.2:   *Square hollow conductors, designation, frequency range, dimensions, electrical data*

Frequency range (H_{10} wave) GHz	Cut-off frequency (H_{10} wave) GHz	Hollow conductor designation 153-IEC	Band	Hollow conductor dimension (cross-section) Width a mm	Height b mm	Tolerance ±mm	Attenuation with copper hollow conductors Frequency GHz	α theoretical dB/m	Maximum dB/m	Max. permissible peak power[1] (between lowest and highest frequency) MW
1.14...1.73	0.908	R 14	L	165.10	82.55	0.33	1.36	0.00522	0.007	12.0–17.0
1.45...2.20	1.158	R 18	D	129.54	64.77	0.26	1.74	0.00749	0.010	7.5–11.0
1.72...2.61	1.375	R 22	–	109.22	54.61	0.22	2.06	0.00970	0.013	5.2–7.5
2.17...3.30	1.737	R 26	S	86.36	43.18	0.17	2.61	0.0138	0.018	3.4–4.8
2.60...3.95	2.080	R 32	A	71.14	34.04	0.14	3.12	0.0189	0.025	2.2–3.2
3.22...4.90	2.579	R 40	G	58.17	29.083	0.12	3.87	0.0249	0.032	1.6–2.2
3.94...5.99	3.155	R 48	C	47.55	22.149	0.095	4.73	0.0355	0.046	0.94–1.32
4.64...7.05	3.714	R 58	J	40.39	20.193	0.081	5.57	0.0431	0.056	0.79–1.0
5.38...8.17	4.285	R 70	H	34.85	15.799	0.070	6.46	0.0576	0.075	0.56–0.71
6.57...9.99	5.260	R 84	T	28.499	12.624	0.057	7.89	0.0794	0.103	0.35–0.46
7.00...11.00	5.790	–	T	25.90	12.95	0.125	8.40	0.0869	0.119	0.33–0.43
8.2...12.5	6.560	R 100	X	22.860	10.160	0.046	9.84	0.110	0.143	0.20–0.29
9.84...15.00	7.873	R 120	M	19.050	9.525	0.038	11.8	0.133	–	0.17–0.23
11.9...18.0	9.490	R 140	P	15.799	7.899	0.031	14.2	0.176	0.230	0.12–0.16
14.5...22.0	11.578	R 180	–	12.954	6.477	0.026	17.4	0.238	–	0.080–0.107
17.6...26.7	14.080	R 220	K	10.668	4.318	0.021	21.1	0.370	0.541	0.043–0.058
21.7...33.0	17.368	R 260	–	8.636	4.318	0.020	26.1	0.435	–	0.034–0.048
26.4...40.0	21.100	R 320	R	7.112	3.556	0.020	31.6	0.583	0.763	0.022–0.031
32.9...50.1	26.350	R 400	Q	5.690	2.845	0.020	39.5	0.815	–	0.014–0.020
39.2...59.6	31.410	R 500	F[2]	4.775	2.388	0.020	47.1	1.060	–	0.011–0.015
49.8...75.8	39.900	R 620	M[2]	3.759	1.880	0.020	59.9	1.52	–	0.0063–0.0090
60.5...91.9	48.400	R 740	E	3.099	1.549	0.020	72.6	2.03	–	0.0042–0.0060
73.8...112.0	59.050	R 900	W	2.540	1.270	0.020	88.6	2.74	–	0.0030–0.0041
92.2...140.0	73.840	R 1200	V[2]	2.032	1.016	0.020	111.0	3.82	–	0.0018–0.0026
114.0...173.0	90.845	R 1400	T	1.651	0.826	–	136.3	5.21	–	0.0012–0.0017

(1) Calculated for the disruptive field strength of E_d = 15 kV/cm, safety factor 2 at sea level
(2) After Hitachi

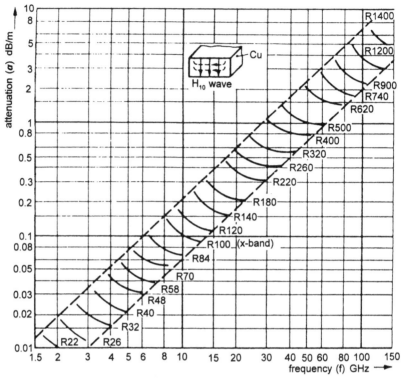

Figure 2.25   *Frequency dependency of attenuation constant of square insulator types; material: copper*

In Figure 2.26, attenuation values are shown for pure energy lines for radio relay systems and for powerful transmitter systems. In comparison with the attenuation values for coaxial cables, as shown in Figure 2.24, the values of the energy line types are substantially lower. Thus, in the case of the HF-1 5/8" cable, attenuation values of about 4 dB/100 m occur at 3 GHz, as opposed to 150 dB/100 m on a conventional RG 58 coaxial cable.

A substantial improvement in attenuation is achieved over the relatively good values indicated, with the use of Siral hollow conductors (Figure 2.27). It is possible, under certain specific preconditions, to extend the frequency ranges shown in Figure 2.25 for the hollow conductors above what is referred to as the single-value limit.

The significance of these single-value limits is that, at higher frequencies, the propagation of electromagnetic waves is no longer clearly defined. It is, for example, possible to have two or more wave types with different propagation rates at one frequency.

Figure 2.28 shows the transmission ranges for the different types of wave conductors. It can be seen from the curves that, for example, with a coaxial cable the single-value transmission range from 0 to 15 GHz is sufficient. In this context, it is

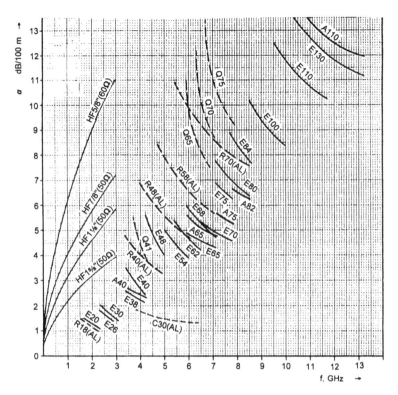

Figure 2.26   *Attenuation constant of high-frequency energy lines*

HF: Flexwell cable, air-space insulated coaxial cable with PE support coil
R: Square conductor, simultaneous transmission of two polarisations
A: ALFORM hollow conductor, oval cross-section, can be drawn off drums
C: Circular hollow conductors, low attenuation; material: Al
E: Flexwell hollow conductors, helical corrugated copper tube, oval
     cross-section, can be drawn off drums

assumed that a cable is involved which is filled with a dielectric of $\varepsilon_r = 2.3$, and for which D + d = 0.8 cm. A square hollow conductor with the wide side a = 0.8 cm has a single-value transmission range from approx. 18.8 GHz to approximately 37.5 GHz. A circular hollow conductor with d = 0.8 cm is 'single-value' in the range from approximately 22 to 28.8 GHz.

Figure 2.29 shows the maximum transferable continuous power rating with different types of hollow conductor. In this situation, a variety of different maximum values are derived, depending on the degree to which the inner wall of the hollow conductor is heated. In the case of transmission of radar pulses, for example, the determinant factor is the dielectric strength of the lines. Compare eqn. 2.34 in the case of coaxial cables. In the case of pulse transmission, the question arises of the

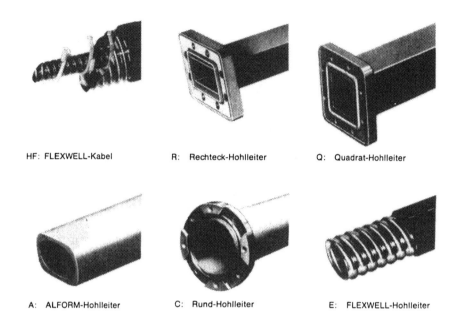

HF: FLEXWELL-Kabel   R: Rechteck-Hohlleiter   Q: Quadrat-Hohlleiter

A: ALFORM-Hohlleiter   C: Rund-Hohlleiter   E: FLEXWELL-Hohlleiter

Figure 2.26   *High-frequency energy lines (continued)*

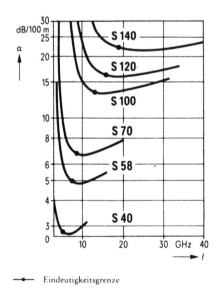

Figure 2.27   *Siral square hollow conductors, transportable on drums, flexible aluminium hollow conductors; wideband hollow conductors capable of being operated in multiple value range*

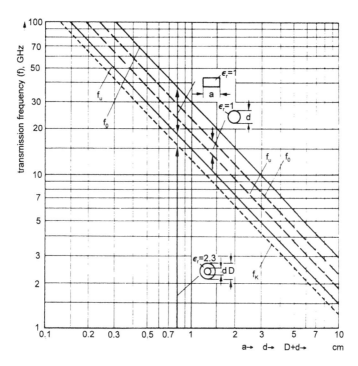

Figure 2.28 *Frequency ranges of coaxial cables, square hollow conductors, and circular hollow conductors*

lower frequency limit: $f_{u\,coax} = 0$; $f_{u\,right}/GHz = 30/2a/cm$;
$f_{u\,round}/GHz = 17.57/d/cm$; upper frequency limit:
$f_{o\,coax}/GHz = 60/\sqrt{\varepsilon_r} \cdot (D+d)/cm$; $f_{o\,right} = 2f_{u\,right}$, $f_{o\,round} = 1.131\,f_{u\,round}$

permissible peak power ratings during the pulse times. The disruptive strength of the hollow conductor is determinant in this context. The theoretical pulse peak power density in square hollow conductors in the case of a disruptive strength value of $E_d = 30$ kV/cm amounts to

$$\frac{P_{sp}}{a\,b}\left[\frac{kW}{cm^2}\right] = 598\sqrt{1 - (f_c/f)^2} \tag{2.35}$$

Here, a and b are the wide and narrow sides of the hollow conductor, respectively, f is the operating frequency, and f_c is the limit frequency, which is associated with the wide side a of the hollow conductor by the equation $f_c = c/2a$. (c is the speed of light in free space.)

The actual radar peak power values are derived on the basis of secondary influences, in percentages according to Figure 2.30.

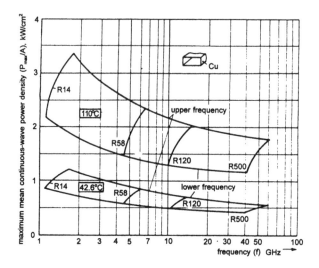

Figure 2.29 *Maximum permissible mean continuous power rating in situations in which wall of hollow conductor is heated to 42.6° and 110°C*

Material: Cu standard hollow conductor

In addition to the standard square hollow conductors shown in Table 2.2, there are other types of hollow conductor with narrow profiles. These are shown in Table 2.3.

In practice, square hollow conductors are also used, as are Flexwell hollow conductors, circular hollow conductors, and elliptical hollow conductors (Figure 2.26). Circular hollow conductors are frequently used as antenna leads, if relatively great lead lengths are required. The attenuation values in these cases are somewhat less than, for example, square hollow conductors with the same cross-section. The single-value transmission range is, however, smaller, with the result that these circular hollow conductors often have to be used in the multiple value range.

Figure 2.31 shows the attenuation values of the groundwave H_{11}, and the attenuation values of the wave types which can cause transmission interference, for a circular hollow conductor with a diameter of 70 mm.

It can be seen that the H_{01} wave of the circular hollow conductor at high frequencies shows extraordinarily low attenuation values. This is the reason for this wave being used for especially low-attenuation transmissions. A disadvantage here, however, is that at low power, wave mode conversions from one type of wave to another may occur.

Figure 2.31 shows attenuation values of waves which are capable of being transmitted in circular hollow conductors in addition to the single-value H_{11} ground-wave, if the single-value ranges shown in Figure 2.28 are exceeded. To be able to compare the attenuation of coaxial cables in cases in which cross-sections are

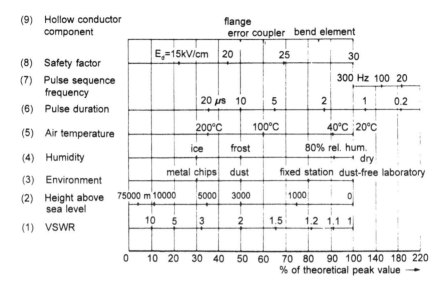

Figure 2.30 *Reduction percentages of theoretical maximum pulse peak power*

approximately the same, the attenuation curve of coax 32/95 is to be incorporated into the picture.

Flexible cables or flexible hollow conductors are of great importance for movable radar systems. In this case, flexible elliptical hollow conductors are frequently used. This type of hollow conductor not only has the advantage of relatively low attenuation and high power transmission capability, but above all the advantage of high flexibility (Figure 2.32), which avoids the high level of expenditure and effort needed in line installation when rigid hollow conductors are used.

Table 2.3   *Narrow-profile hollow conductors*

Frequency range h/GHz	Limit frequency/GHz	Hollow conductor designations	Band	Hollow conductor internal dimensions		
				width mm	height mm	tolerance mm
1.72... 2.61	1.375	F 22	–	109.22	13.1	0.11
2.17... 3.30	1.737	F 26	–	86.36	10.4	0.09
2.60... 3.95	2.080	F 32	S	72.14	8.6	0.07
3.22... 4.90	2.579	F 40	A	58.17	7.0	0.06
3.94... 5.99	3.155	F 48	G	47.55	5.7	0.05
4.64... 7.06	3.714	F 58	C	40.39	5.0	0.04
5.38... 8.18	4.285	F 70	J	34.85	5.0	0.035
6.57... 9.99	5.260	F 84	H	28.499	5.0	0.03
8.20...12.50	6.560	F 100	X	22.86	5.0	0.025

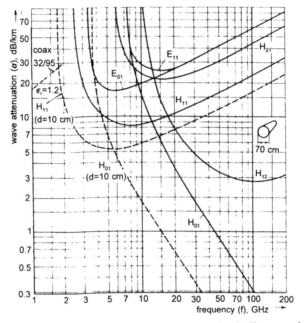

Figure 2.31   *Wave attenuation with 70 cm circular hollow conductor*

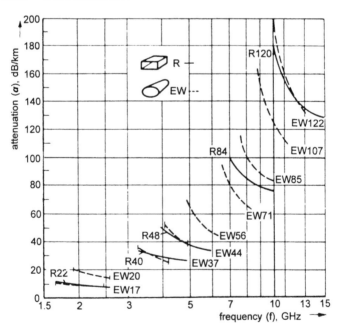

Figure 2.32   *Comparison of groundwave attenuation of square and elliptical hollow conductors*

Chapter 3

Striplines

W. Janssen

Striplines are being used increasingly as conducting elements in circuits in the frequency range from about 300 MHz to about 30 GHz, the reason for this being the trend towards circuit miniaturisation. The technology of striplines has today improved to the point where comparatively cheap microwave circuits can be made, capable of good replication, which can completely fulfil the demands made on them. To a large extent, in the microwave range below 30 GHz, hollow conductors or other types of conductors only continue to be used in cases in which high transmission capacities are required, or in which stringent requirements are placed on the low attenuation properties of the lines.

One characteristic feature of stripline technology is that the circuits can be arranged in one plane of a common carrier board, called the substrate. This fact leads to a high integration capability with microwave circuits. Semiconductor components can be combined with stripline elements to form microwave integrated circuits (MIC). Figure 3.1 shows a series of the most important striplines. In most cases, asymmetrical striplines, known as microstrips, are preferred to other striplines for use as conductor elements in microwave circuits because of their simple structure.

Passive components can be created very easily with the help of stripline technology. Thus, for example, capacitative and inductive reactances can be created very easily through unloaded lines, the ends of which can be brought to the correct length by mechanical tuning; in other words, grinding off.

3.1 Striplines and their properties

3.1.1 Structure of microstrip

In microstrip, the dielectric carrier material is covered on one side entirely by a conductive coating, with the line structure applied to the other. Copper, gold, and, at lower frequencies, aluminium are the conductor materials of preference because of their good conductive properties and their resistance to corrosion.

Figure 3.2 shows the composition of a microstrip, in a schematic representation. The waveform of the conductor wave corresponds approximately to that of a TEM wave.

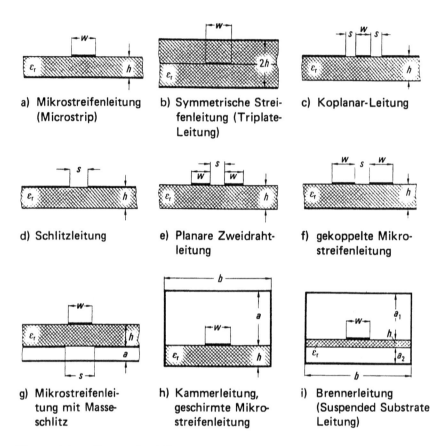

Figure 3.1   *Types of stripline*

The greater the dielectric constant ε_r of the substrate, the more the total field is drawn into the interior of the dielectric, and therefore the more the conducting properties of the conductor are improved. Table 3.1 shows the most important

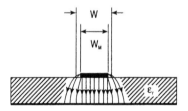

Figure 3.2   *Field pattern with microstrip line*

W = effective conductor path width
W_M = true conductor path width

Table 3.1   *Substrate parameters*

Substrate	Copper laminated on both sides	Copper cover (t/%m)	Substrate thickness (h/mm)	Dielectric constant ε_r	Frequency GHz (typical)
Glass-fibre-reinforced epoxy resin (FR 4)	Yes	35/70	1.58±0.18	5±0.5	<1
Glass-fibre-reinforced Teflon laminate	Yes	35/70	1.56	2.5	3
Polyguide Teflon	Yes	35/70	1.58	2.32	5
Cu Flon	Yes	35/70	1.58	2.06	4
RT/duroid 5870 (micro-glass-fibre reinforced)	Yes	17/35/70	1.6...0.8	2.35±0.02	1...10
RT/duroid 5880 (micro-glass-fibre reinforced)	Yes	17/35/70	0.8 0.58 0.25 (0.12)	2.23±0.02 2.20±0.02	10 15 20
RT/duroid 6010 (ceramic PTFE)	Yes	35	1.27±0.05	10.5±0.25	15
Aluminium ceramics Al$_2$O$_3$ (99.5% purity)	No	2...10	0.635 0.508 0.381 0.254	9.7±0.05 9.7±0.05	8...20 15...25 15...25 20...30
Epsilan 10 (plastified)	No		0.635	9.8	15
Ferrite	No	2...10	0.508	12...16	10

substrates used in practice. Al$_2$O$_3$ substrates with a thickness of 0.635 or 0.508 mm are used relatively often at frequencies of 10 GHz. Such substrates are available commercially in outer dimensions of, among others, 1×1 inch (25.4 × 25.4 mm), or 1×2 inch (25.4 × 50.8 mm). The thickness h of the substrates in this case varies depending on the frequency range used. Substrate thicknesses h as per Table 3.2 are frequently chosen for circuits with the conventional and widespread aluminium oxide substrates ($\varepsilon_r = 9...10$) for the different frequency bands.

Table 3.2   *Substrate thicknesses*

Frequency range GHz	h mil	h mm
8–15	25	0.635
10–18	20	0.508
13–22	15	0.381
18–35	10	0.254

Conductor paths on ceramic substrates (Al_2O_3) are produced by evaporation coating or galvanisation coating. In these cases, gold is often galvanised onto a thin copper coating, as the conductor material. Gold is well-suited as a conductor material because of its good corrosion resistance properties. In the case of double-laminated substrates with copper coatings, conventional etching technology is applied for the manufacture of the conductor paths.

3.1.2 Transmission characteristics of microstrip

Wave impedance, wavelength, and attenuation constant characterise the transmission properties of lines. Figure 3.3 shows the wave impedances of microstrips for various different values of ε_r, as a function of the line width and substrate thickness. To calculate these resistance values, Hammerstadt formulas are used.

The wave impedance of microstrip with air as the dielectric ($\varepsilon_r = 1$) is obtained thus:

$$Z_0/\Omega = 60 \ln (8h/w + w/4h) \text{ for } \frac{w}{h} < 1 \tag{3.1}$$

$$Z_0/\Omega = 120\pi/(w/h + 1.393 + 0.667 \ln(w/h + 1.444)) \text{ for } \frac{w}{h} \geq 1 \tag{3.2}$$

The wave impedance Z_L of a microstrip, the substrate of which has the dielectric constant ε_r, is derived from the formula

$$Z_L = Z_0/\sqrt{\varepsilon_{eff}} \tag{3.3}$$

The effective dielectric constant ε_{eff} is not identical to ε_r because of the layered structure of the media. The following relationships are derived for ε_{eff}:

$$\varepsilon_{eff} = \frac{1}{2} (\varepsilon_r + 1) + \frac{1}{2} (\varepsilon_r - 1) F \tag{3.4}$$

The value of the factor F in this case is

$$F = (1 + 12\ h/w)^{-1/2} + 0.04\ (1 - w/h)^2 \quad \text{for } w/h \leq 1 \tag{3.5}$$

$$F = (1 + 12\ h/w)^{-1/2} \qquad\qquad \text{for } w/h > 1 \tag{3.6}$$

The strip widths w in eqns. 3.1 – 3.6 are what are known as effective line values. The actual conductor paths w_M applied, whether electrically or by etching technology, are related to the thickness t of the conductor layer with the effective conductor widths (Figure 3.2) as follows:

$$W = w_M + \frac{t}{\pi}\ (1 + \ln\ (2h/t))\ \text{for } w_M/h \geq \frac{1}{2\pi} \tag{3.7}$$

The conductor width is calculated by means of the finite thickness of the conductive layer, with a correction which is generally only slight. For example, with $t = 10\ \mu m$, $h = 0.635$ mm at $Z_L = 50\ \Omega$, $\varepsilon_r = 9.7$, the error is $w = w_M + 0.02\ w_M$; $w = 0.60$ mm; $w_M = 0.59$ mm.

The relationship between the stripline threshold length and the wavelength in free space is designated the shortening factor F_v:

$$F_v = \lambda_\varepsilon/\lambda_0 = 1/\sqrt{\varepsilon_{\text{eff}}}. \tag{3.8}$$

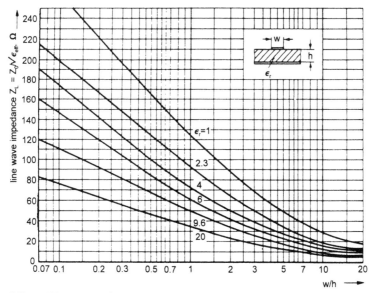

Figure 3.3   *Wave impedance of microstrip, calculated according to Hammerstadt (eqn. 3.1) for $f \ll f_p = Z_L/2\mu h$*

This factor is represented in Figure 3.4. The shortening factor is increasingly dependent on frequency, especially in the upper frequency ranges. To eliminate the interfering influences of the frequency dependency as far as possible, upper frequency ranges should be used.

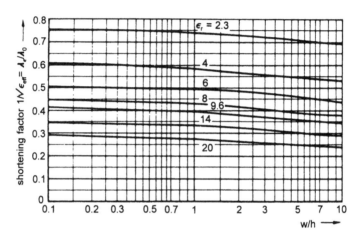

Figure 3.4   *Shortening factor of microstrip, calculated according to Hammer-stadt for f<f$_p$ = Z$_0$/2µh*

From a defined upper frequency limit f$_{cob}$ onwards, the frequency dependency of the effective dielectric constant takes effect as follows:

$$f_{cob}/GHz = 0.3 \ (Z_0/(h(\varepsilon_r - 1)^{1/2}))^{1/2} \tag{3.9}$$

where Z$_0$ is calculated in Ω and h in cm. An estimate of the error limits can be determined by means of the following formula:

$$\varepsilon_{eff}(f) = \varepsilon_r - \left(\frac{\varepsilon_r - \varepsilon_{eff}}{1 + G(f/f_p)^2} \right) \tag{3.10}$$

where   G $= 0.6 + 0.009 \ Z_0/\Omega$

f $=$ operating frequency

f$_p$ $= Z_L/2h\mu_0$ with Z$_L$ in Ω, h in cm

μ_0 $= 4\pi \cdot 10^{-9}$ H/cm

ε_{eff} $=$ effective dielectric constant according to eqn. 3.4.

For frequency values $f \geq f_{cob}$, the corrected values $\varepsilon_{eff}(f)$ are used in eqns. 3.3 and 3.8 instead of ε_{eff}. For example, with Z$_L$ = 50 Ω, ε_r = 9.7, h = 0.635 mm; f$_p$ = 31.3 GHz, with f = 4.9 GHz, $\varepsilon_{eff}(f)$ = 6.63 instead of ε_{eff} = 6.56. The difference in the corrected dielectric constants below, and in the vicinity of, f_{cob} is relatively low, as this example shows.

The cause of this unwanted frequency dependency can be seen to be the occurrence of surface waves. As frequencies rise, these waves are excited in particular with relatively broad conductor paths and relatively large chip thicknesses.

Line losses in the case of striplines are substantially greater than with hollow conductors. This disadvantage can be compensated for by short line lengths. Line loss attenuations are derived according to eqn. 3.11. These values agree relatively well with practical measurement results (Figure 3.5):

$$\alpha_c / \frac{dB}{cm} = \frac{200}{\ln 10} \frac{R_s/Z_L}{w/mm} = 5.46 \sqrt{10^6 \frac{f/GHz}{\kappa/1\Lambda\Omega m} \frac{1}{w/mm} \frac{\sqrt{\varepsilon_{eff}}}{Z_0/\Omega}} \tag{3.11}$$

The dielectric losses, which in general lie below 10 % of the losses owing to heat (I^2R), amount to

$$\alpha_c / \frac{dB}{cm} = 272.9 \ (1-1/\varepsilon_{eff}) \ \tan\delta/(1-1/\varepsilon_r) \ \lambda_\varepsilon/mm. \tag{3.12}$$

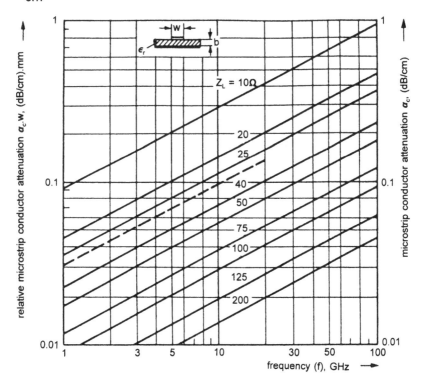

Figure 3.5   *Microstrip conductor attenuation; material: copper, loss angle of substrate $\tan \delta \ll 1, f < f_p$ [1, 3]*

3.1.3 Symmetrical stripline (triplate stripline)

In the case of a symmetrical stripline, the effective dielectric constant ε_{eff} and the relative dielectric constant ε_r are equally large. Accordingly, as with coaxial lines, the shortening factor does not depend on the conductor dimensions and we derive:

$$\lambda_\varepsilon/\lambda_0 = 1/\sqrt{\varepsilon_{eff}} = 1/\sqrt{\varepsilon_r}. \tag{3.13}$$

Figure 3.6 shows the wave impedance Z_L as a function of the line dimensions w/h. The wave impedance of the symmetrical striplines can be calculated for the conductor density $t = 0$, from the following equations:

$$Z_L \sqrt{\varepsilon_r}/\Omega = 60 \ln \left(2 \coth \left(\pi w/4b\right)\right) \text{ for } Z_L \sqrt{\varepsilon_r} > 30 \, \pi. \tag{3.14}$$

$$Z_L \sqrt{\varepsilon_r}/\Omega = 15 \, \pi^2/\ln \left(2e^{\pi w/2b}\right) \text{ for } Z_L \sqrt{\varepsilon_r} < 30 \, \pi. \tag{3.15}$$

A symmetrical stripline configuration can be derived, for example, from two microstrips which are laid back to back. To restrict the large number of types used, a series of symmetrical striplines has been standardised, Table 3.3. The wave impedance of these lines is 50 Ω. Internal and external conductors in this context

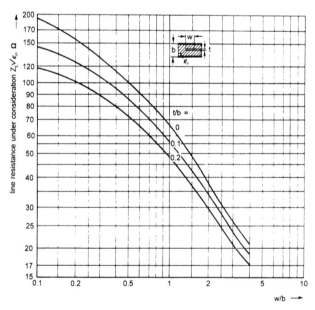

Figure 3.6   *Wave impedance of symmetrical stripline according to eqns. 3.14 and 3.15*

Table 3.3   *Standardised symmetrical striplines*

Designation	Chip spacing b mm	Conductor thickness d mm	Conductor width w mm	Minimum spacing between conductor sides mm	Upper frequency limit GHz
MPC-062-2	3.15	0.07	2.108±038	11.3	7.5
MPC-125-2	6.53	0.07	4.623±076	16.9	5.0
MPC-187-2	9.5	0.07	7.112±102	22.6	3.6
MPC-250-2	12.7	0.07	9.652±127	28.3	2.8

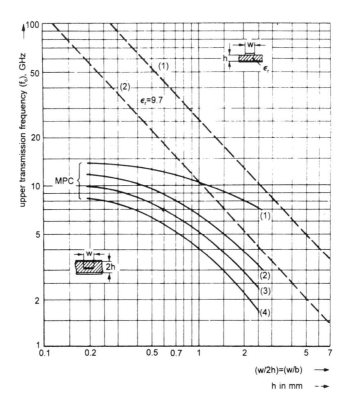

Figure 3.7   *Frequency ranges with microstrip (broken curve) and symmetrical stripline (unbroken curve)*

(1)   Theoretical upper transmission limit with microstrip
($\varepsilon_r = 9.7$) $f_{ob}/GHz = 75/\sqrt{\varepsilon_r - 1}$ h/mm
Technical frequency limit (dispersion $\Delta\lambda/\lambda = 5\%$)

(1)-(4)   Upper practical frequency limit with symmetrical stripline with MPC 062, 125, 187, 250, respectively

consist of a copper layer 0.07 mm thick. The dielectric filler material has a dielectric constant of $\varepsilon_r = 2.55$.

Figure 3.7 shows the frequency range of the various types of striplines as a function of the conductor width w/b. Curves (1) – (4) in Figure 3.7 show how, in this case, the upper frequency limit decreases as the conductor width increases. The values of upper-frequency width of microstrip are compared in these curves. The microstrip has a theoretical maximum transmission limit, which can be calculated according to the formula:

$$f_c/\text{GHz} = 75/\sqrt{\varepsilon_r - 1}\ h/\text{mm} \tag{3.16}$$

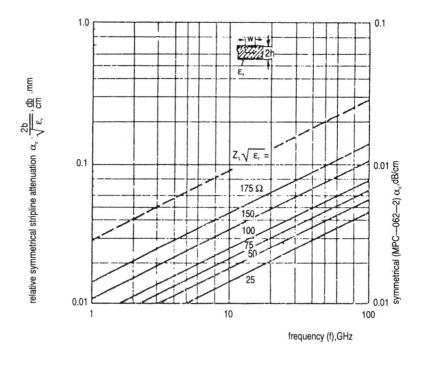

Figure 3.8   *Attenuation constant with symmetrical stripline*

Material: copper; related line attenuation $a_c\ 2b/\sqrt{\varepsilon_r}$ line attenuation in dB/cm at MPC-062

This maximum transmission limit f_C with microstrip indicates the limit for the occurrence of line-dependent surface waves, and only indirectly has anything to do with the values from eqns. 3.9 and 3.10, which are only of theoretical importance.

Figure 3.8 represents the attenuation of symmetrical striplines as a function of frequency.

3.1.4 Slotlines

Slotlines are used for mixers, filters etc., particularly in the higher frequency range. In cases in which wave impedances of the lines $Z_L > 100\,\Omega$ are of advantage, slotlines are used. The microstrip has relatively high power losses with high wave impedances. This applies in particular in the upper frequency ranges ($f > 10$ GHz).

Figure 3.9 shows the wave impedances of slotlines as a function of W/A. In this situation, A is the width of the entire line system. In Figure 3.9 it is taken as a precondition, by way of simplification, that the line waves are pure transverse waves

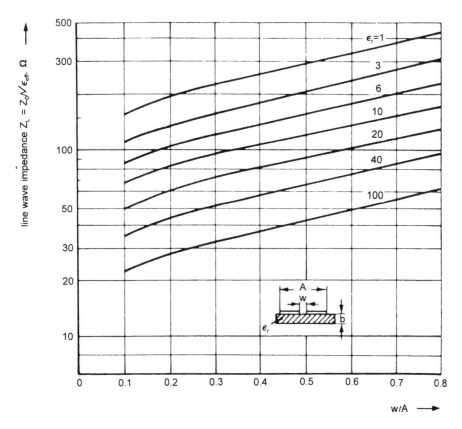

Figure 3.9   *Wave impedance of slotline for infinitely large values of* b

parameter: dielectric constant ε_r

and the substrate thickness is infinitely great. With these preconditions, the wave impedance amounts to

$$Z_L = Z_0/\sqrt{\varepsilon_{eff}} = Z_0/\sqrt{(\varepsilon_r + 1)/2} \tag{3.17}$$

As the value of Z_0, we obtain

$$Z_0/\Omega = 120 \ln\left(2\frac{\sqrt{A/w} + 1}{\sqrt{A/w} - 1}\right) \text{ for } Z_0 \geq 188.49 \ \Omega \tag{3.18}$$

$$Z_0/\Omega = 30 \ \pi^2/\ln(2 \ \sqrt{A/w}) \quad \text{for } Z_0 \leq 188.49 \ \Omega \tag{3.19}$$

Figure 3.10 shows the shortening factor:

$$Fv = \frac{\lambda_\varepsilon}{\lambda_0} = \frac{1}{\sqrt{\varepsilon_{eff}}} = \sqrt{\frac{2}{\varepsilon_{eff} + 1}} \tag{3.20}$$

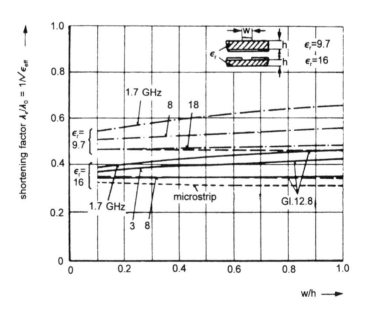

Figure 3.10   *Comparison of shortening factor of microstrip and slotline*

parameter: frequency f

As an example, in Figure 3.10 the relationship of shortening factor of the quotient line spacing interval w and substrate thickness h is shown for different frequency parameters. To provide a comparison, the shortening factor of a microstrip is also plotted. The value $1/\sqrt{\varepsilon_{eff}}$ is substantially smaller with the microstrip than with the slotline.

3.1.5 Coplanar lines

The coplanar wave line, as shown in Figure 3.11, consists of three line strips, which, in a similar manner to the microstrips, are intended to be located on a dielectric which is as loss-free as possible. The electrical fields in this case run from the outer,

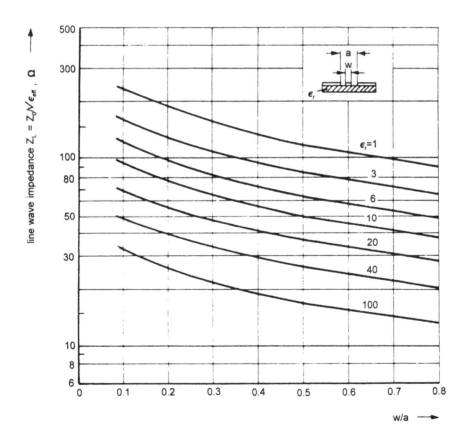

Figure 3.11   *Wave impedance of coplanar line*

parameter: dielectric constant, substrate thickness infinite

somewhat broader line strips, to the line strips located on the inside. The fields are denser in the area of the inside conductor. The outer conductor paths in most cases do not exceed the values of 5–10 mm, since the characteristic line data is only slightly influenced by wider paths.

In Figure 3.11, the wave impedance is shown as a function of w/a. In the frequency range of between 10 and 20 GHz, the conventional 99.6 % Al2O3 ceramic substrate of 0.508 or 0.635 mm is often chosen, as it is for other striplines.

An advantage with coplanar lines as with slotlines is the fact that hybrid elements such as diodes, transistors, etc. can be used in the circuit without having to drill through the material, which is mostly very hard and brittle. Parallel circuits of hybrid elements can be created more easily with these lines than, for example, with microstrips.

With coplanar lines, the thickness of the substrate is largely of no consequence, which means that this kind of conductor can also be laid on dielectric material which serves not only as the carrier material for the lines, but can also perform other design and structural functions.

3.1.6 Microstrip lines with earth slot

Figure 3.12 shows the wave impedance of a microstrip with an earth slot, as a function of w/h. This type of line is used with stripline circuits in which it is intended to create direction-dependent power couplings or power distribution arrangements. The substrate in this case is accommodated in a housing in such a way that the earthing side of the stripline is located at a defined distance a from the housing. In Figure 3.12, the carrier material is assumed to be aluminium ceramic (ε_r= 9.7; h = 0.635 mm).

To avoid resonances between the underside of the substrate and the housing, certain specific dimensional specifications must be respected, depending on the frequency.

Example: It is intended to design the dimensions of a microstrip according to Figure 3.13. Two dipole antenna (input resistance R_{DIP} = 50 Ω) are intended to be fed via a Wilkinson power distributor at 5 GHz, free of reflection, with the same power values. All connection lines are designed in 50 Ω technology. Polyguide is used as the substrate: ε_r = 2.32; h = 1.58 mm.

Solution: (i)   For the wavelength in air
λ_o/cm = 30/f/GHz; λ_o = 6 cm

(ii)   Matching is applied at the branching point (3) if a resultant resistance of $R_{(3)}$ = 50 Ω occurs there. The input resistance $R_{(3)-(1)}$ of the line to the dipole (1) must then be 100 Ω and the input resistance to the dipole (2) $R_{(3)-(2)}$ must likewise be 100 Ω.

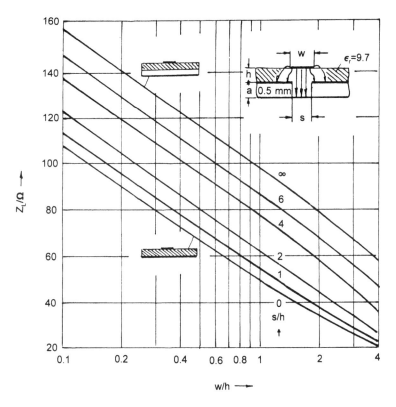

Figure 3.12 *Wave impedance of microstrip with earth slot [22]*

(iii) At point (1) of the line Z_L, the resistance $R_A = 50\ \Omega$, and likewise at point (2) of the line. This resistance must be transformed into a resistance of $100\ \Omega$ over a line Z_{LT} of the length $\lambda_\varepsilon/4$. This is possible with $Z_{LT} = \sqrt{R_{31} \cdot R_A} = 70.7\ \Omega$. Both lines accordingly have the wave impedance $Z_{LT} = 70.7\ \Omega$.

(iv) Width w_0 of the 50 Ω line. From Figure 3.3, for $t = 0$: $w_0/h = 2.9$; $w_0 = 4.6$ mm. Width w_{LT} of the 70.7 Ω line: $w_{LT}/h = 1.7$; $w_{LT} = 2.7$ mm.

(v) Length of the transformation line $l = \lambda_\varepsilon/4$. With Z_{LT}, according to Figure 3.4, for $w_{LT}/h = 1.7$; $F_v = 0.73$; $\lambda_\varepsilon = \lambda_0 F_v = 43.8$ mm. $l = 11.0$ mm.

(vi) All 50 Ω lines can be of any desired length (subject to freedom from loss).

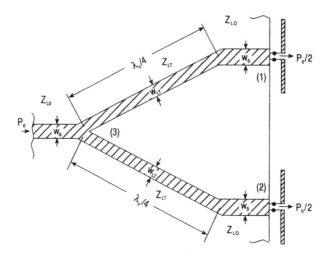

Figure 3.13 *Wilkinson power distributor*

3.2 Stripline circuits

3.2.1 Basic circuits

Depending on their actual significance, striplines serve to provide the connection between components and circuits. In other words, they do not serve to provide an electrical transfer over any very substantial distance (typical: $l < 10$ cm), because of their relatively high attenuation.

Complete circuit groups can be compiled from line elements. In view of the fact that the line dimensions of striplines usually lie in the order of magnitude of the wavelength, short-circuited or unloaded lines can be used to represent passive circuit elements, such as capacitances and inductances. In Figure 3.14, in addition to these, a series of examples in stripline technology are illustrated. A line section of length l, in the transient condition, can be represented formally as a T-element or π-element.

In Figure 3.14a, L and C of the equivalent circuits are given for stripline (in particular for symmetrical and for asymmetrical striplines). For small line lengths ($l \ll \lambda_\varepsilon$), relatively simple relationships are derived. With matching circuits, spur lines are frequently used which are applied in T-shape on the main line. Figures 3.14b and c show how any desired susceptance values can be transformed along the line. With microstrip, it should be borne in mind that the effective point at which the line

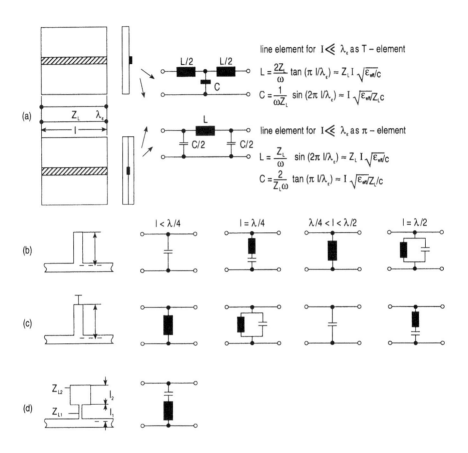

The following equations accompany figure (a):

line element for $I \ll \lambda_e$ as T – element

$$L = \frac{2Z_L}{\omega} \tan(\pi\, I/\lambda_e) \approx Z_L\, I\, \sqrt{\varepsilon_{eff}}/c$$

$$C = \frac{1}{\omega Z_L} \sin(2\pi\, I/\lambda_e) \approx I\, \sqrt{\varepsilon_{eff}}/Z_L c$$

line element for $I \ll \lambda_e$ as π – element

$$L = \frac{Z_L}{\omega} \sin(2\pi\, I/\lambda_e) \approx Z_L\, I\, \sqrt{\varepsilon_{eff}}/c$$

$$C = \frac{2}{Z_L \omega} \tan(\pi\, I/\lambda_e) \approx I\, \sqrt{\varepsilon_{eff}}/Z_L c$$

Figure 3.14 *Representation of inductive and capacitative reactances through lines*

(a) See to right of illustration
(b) Unloaded spur line:
   $X = Z_L \cot(2\pi/\lambda_e)$
(c) Short-circuit spur line:
   $X = Z_L \tan(2\pi l/\lambda_e)$
(d) Line series resonant circuits:
   $L \approx Z_{L1} l_1 \sqrt{\varepsilon_{eff1}}/C$; $C \approx l_2 \sqrt{\varepsilon_{eff2}}/Z_{L2} \cdot c$
   c = velocity of light

relates to the line equivalent circuit cannot be calculated exactly from the midpoint of the stripline path. In addition to this, because of the stray capacitances with unloaded line ends, the effective line length l is greater than the mechanically measurable length by Δl:

$$\Delta l = h \, 0.41 \, \frac{\varepsilon_{eff} + 0.3}{\varepsilon_{eff} - 0.26} \cdot \frac{w/h + 0.26}{w/h + 0.81} \tag{3.21}$$

Series and parallel resonance circuits can be set up according to Figure 3.14, with line elements l/2 long, but also with the aid of short line sections (Figure 3.14d).

Coupling capacitances in the series branch of lines are frequently obtained by means of capacitance diodes. Other possibilities for providing coupling capacitances for line ends are shown in Figure 3.15. By causing conductor paths to overlap (Figure 3.15b), which are electrically separated by very thin electrically insulating materials (e.g. silicon oxide), it is often possible to create coupling capacitances of sufficient value:

$$C_{coupler}/pF = 8.84 \; 10^{-3} \, \varepsilon_r \, \frac{w l/mm^2}{d/mm} \tag{3.22}$$

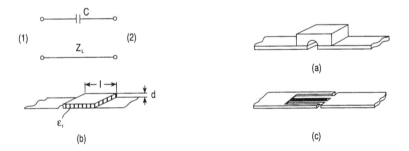

Figure 3.15 *Coupler capacitances in microstrip technology*

    (a)   Capacitative pn transition
    (b)   Coupler capacitance, electrical separation of conductor paths by means of
           insulating layer applied by evaporation coating (e.g. SiO2), overlay structure
    (c)   Comb-type toothed arrangement of line ends

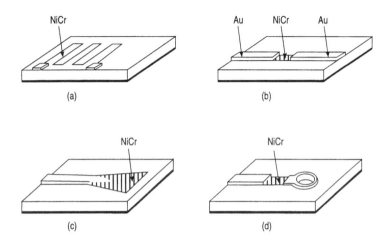

Figure 3.16 *Integrated resistors*

    (a)   Meandering resistance path (e.g. NiCr in Al$_2$O$_3$ technology)
    (b)   Microwave integrated series resistor
    (c)   Wideband reflection-free line termination
    (d)   Line termination with contacts for conductor path and motherboard

By means of line ends which engage with one another in comb-fashion (Figure 3.15c), capacitance values can be obtained, depending on the design, of the order of several pF.

Resistances can be created by spraying on nickel–chrome layers. Thus, for example, in the microwave range, with Al$_2$O$_3$ substrates, it is possible to create resistances of between 0 and about 200 Ω (Figure 3.16). In the case of direct currents, or at low frequencies, resistances of up to about 200 kΩ can be achieved by means of meandering patterns of thin nickel–chrome lines.

Extensive circuits can be formed on relatively small substrates (2.54 × 2.54 cm). These circuits, known as MICs or microwave integrated circuits, may consist of complete amplifier units, mixer units, modulator units, filter units, and so on. In this case, where linear and nonlinear components (line and semiconductor components) are combined, we speak of hybrid circuits.

Figure 3.17 shows a mixer in MIC technology. In this situation, coplanar lines, slotlines, and microstrips are assembled on a common substrate chip to form one single-function unit. The substrate in this case is populated on both sides with the different line elements.

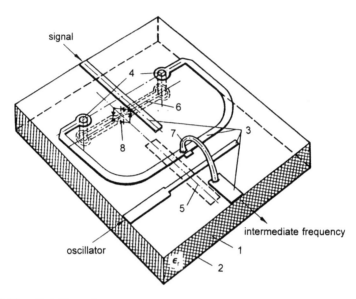

Figure 3.17   *Stripline mixer*

Push-pull mixer, formed from
(1)   Substrate Al$_2$O$_3$ ceramics
(2)   Earthing side, conductor side with copper-gold overlay
(3)   Microstrip, slotline
(4)   Contact pin, from top side to bottom side, from microstrip to coplanar line
(5)   Slotline, with double coupling to microstrip
(6)   Coplanar line
(7)   Contact bar
(8)   Push-pull mixer diodes

Slot lines which are located on the opposite side to the microstrips, can, as shown in Figure 3.17, transfer energy from one type of conductor to another simply by causing their lines to be crossed. The signal voltage, coupled in two equal parts via a slotline, is mixed in the coplanar line section with the oscillator voltage. The oscillator voltage is coupled in via a microstrip, and transferred again on a slotline to the two coplanar lines. The decoupled intermediate-frequency voltages are decoupled via a microstrip.

3.2.2 Filter circuits

3.2.2.1 Lowpass
A lowpass filter can be formed by connecting short line elements one after another. In this context, the line resistors Z_L of the lines must have alternating high and low values.

Figure 3.14a shows that a line can be represented by a T-element or a π-element. In the case of a line element with high impedance Z_L, provided that the line lengths are small, the ratio is $Z_L = \sqrt{L/C}$ i.e. the inductance outweighs the capacitance C. This line element can be represented by way of approximation, in the transmission range under consideration, as an inductive element (T-element), since a lower impedance is formed by L and a higher one by C. Conversely, in cases of low line impedances, the capacitative behaviour predominates.

Figure 3.18a shows a stripline lowpass filter as a series connection of π-elements and T-elements. By means of this equivalent circuit, the lowpass behaviour of a stripline filter can be determined relatively precisely. The stray capacitances shown in Figure 3.18a are small in comparison with the line capacitances and are therefore in many cases not taken into consideration when making calculations. The values of L and C of the lines can be derived from the representations given in Figure 3.14a.

From Figures 3.18b and c, simple equivalent circuits of lowpass filters can be obtained for line length l.

For dimensioning of components for filters of this kind, we derive

$$\frac{l_2/\lambda_{\varepsilon 2}}{l_1/\lambda_{\varepsilon 1}} = \frac{Z_{L2}}{Z_{L1}} \frac{1 - Z_{L1}^2/Z_L^2}{Z_{L2}^2/Z_L^2 - 1} \tag{3.23}$$

For the situation with $l_2\lambda_{\varepsilon 1}/l_1\lambda_{\varepsilon 2} = 1$, we obtain the solution

$$Z_{L1} \cdot Z_{L2} = Z_L^2. \tag{3.24}$$

By variation of the line lengths l_1 and l_2, and by variation of the line impedances (L, C) in the filter chain as shown in Figure 3.18a, lowpass curves can be attained according to the requirements imposed. The dimensions of the conductor paths are derived from the values of Z_L and λi, depending on whether the circuit in question is a microstrip, an asymmetrical stripline, etc. A lowpass filter circuit with relatively steep attenuation flanks is shown in Figure 3.18d. With this type of filter, attenuation poles are attained with

$$\omega_{p1,2} = 1/\sqrt{L_{2,4}C_{2,4}} \tag{3.25}$$

This circuit is described as a Cauer lowpass filter.

3.2.2.2 Bandpass

Figure 3.19 shows a series of bandpass filters in stripline technology. Depending on the purpose, different requirements are made on the frequency filters. This results in the large number of filter types which exist, of which only a few representatives of the most important types are shown here. Characteristic transmission values are determined by the selection of the midfrequency, the bandwidth, the steepness of the filter curve, and the attenuation in the frequency range in question.

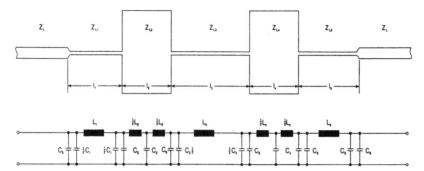

(a) Microstrip lowpass, layout and equivalent circuit, stray capacitances C_S at line transition points

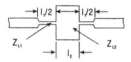

(b) Lowpass filter and equivalent circuit, T-element

$X_L \approx Z_{L1} \, 2\pi \, (l_1/\lambda_{\epsilon 1}) \, (1-Z_L^2/Z_{L1}^2);$

$B_c \approx (2\pi/Z_{L2}) \, (l_2/\lambda_{\epsilon 2}) \, (1-Z_{L2}^2/Z_L^2);$

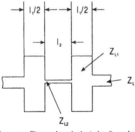

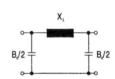

(c) Lowpass filter and equivalent circuit, π-element

$X_L \approx Z_{L2} \, 2\pi \, (l_2/\lambda_{\epsilon 2}) \, (1-Z_L^2/Z_{L2}^2);$

$B_c \approx (2\pi/Z_{L1}) \, (l_1/\lambda_{\epsilon 1}) \, (1-Z_{L1}^2/Z_L^2);$

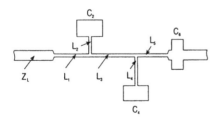

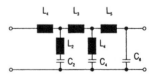

(d) Lowpass filter with equivalent circuit

Figure 3.18 *Stripline lowpass filter circuits*

In the examples shown in Figure 3.19, the designs are essentially from microstrip and symmetrical stripline technology. In this situation, filter types with lateral short-circuit limitation are well suited (Figures 3.19d, g, h), especially for symmetrical stripline designs.

The filter type shown in Figure 3.19c is capable of being implemented in both types of technology, and can be used for all possible variations of lengths l and line

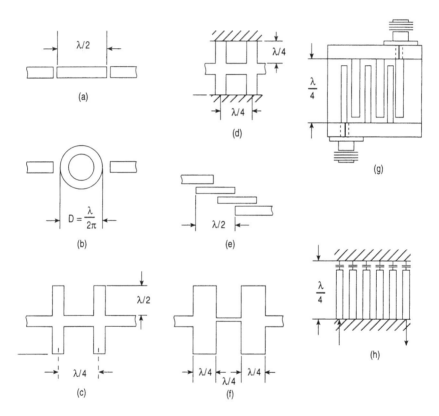

Figure 3.19 *Stripline bandpass filters*

- (a) Coupled $\lambda/2$ line
- (b) Coupled ring line
- (c) Bandpass with unloaded spur lines
- (d) Bandpass with short-circuit spur lines
- (e) Bandpass with coupled $\lambda/2$ line resonators
- (f) Wide bandpass filters
- (g) Interdigital filters, six-finger with steep filter flank
- (h) Comb-line filter, six-finger, capable of being capacitatively matched

impedances Z_L. In practice, the filter designs of Figures 19e, g and h are frequently chosen.

For narrowband filters (bandwidth about 10 to 15 %), the simple design of the bandpass formats shown in Figure 3.19e are well suited. In this case, the striplines, $\lambda/2$ long, which are coupled to one another with a mutual interval spacing of specific size, provide the filter effect which is required.

Interdigital and comb-line bandpass filters (Figures 3.19g and h) are used in the frequency ranges from about 100 MHz to 30 GHz. To guarantee the load-carrying capacity of the internal conductor in symmetrical striplines with air as insulation ($\varepsilon_r = 1$), they are secured permanently to the edge of the line. The conductors in this case frequently have round or even rectangular cross-sections. In the example shown in Figure 3.19h, variable capacitances are applied for the correctional matching of the filter. These capacitances are formed by screw-in conductor surfaces, which are separated from the ends of the comb-type coupling lines by a variable interval.

Figure 3.19f shows part of a wideband filter, which belongs to a chain of high-ohmic and low-ohmic $\lambda/2$ lines connected alternately, one behind another.

3.2.3 Stripline directional couplers

Directional couplers are used in all situations in which the decoupling of the wave components that run backwards and forwards is required, and when power gains are produced. The directional coupler has a range of potential applications. It can be used, for example, as a power distributor, frequency diverter, switch, mixer, and in decoupling circuits in matching and reflection measurement circuits.

3.2.3.1 Hybrid couplers (branchline couplers)

Figure 3.20 shows an array of different hybrid couplers. These function as power distributors with directional detector effect; for example, the power P_1 fed in at port (1) is transferred to port (2) and to port (3). The couplers are frequently dimensioned in such a way that the power is distributed into two portions of equal size: $P_2 = P_3$. This coupler is, in this case, referred to as a 3 dB coupler (the level of P_2 and P_3 is 3 dB below the level of P_1).

Port (4) is decoupled in the case of Figure 3.20 ($P_4 = 0$). The voltage fractions at port (2) and port (3) have a mutual phase difference, which in the case of the 3 dB couplers is 90°. In the case of the 3 dB square hybrid, and the ring hybrid (Figure 3.20a, c), the wave impedances of the longitudinal branches amount to $Z_L/\sqrt{2}$ (= 35.3 Ω at $Z_L = 50\ \Omega$). The wave impedances of the connection lines amount to Z_L. The line lengths in each case amount to a quarter of the wavelength. A type of coupler composed of square elements (H-line coupler) is shown in Figure 3.20b.

If the input (e.g. port (1)) of the branch-line coupler is matched to Z_L at the midfrequency, the following equations are derived for the case of the complete decoupling operation (e.g. of port (4)).

With square hybrids (Figure 3.20a):

$$P_2/P_3 = (Z_L/Z_{L1})^2; \quad (Z_L/Z_{L2})^2 = 1 + (Z_L/Z_{L1})^2 \tag{3.26}$$

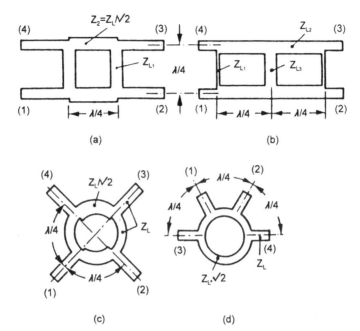

Figure 3.20   *Stripline directional coupler*

     (a)   Square coupler (square hybrid); power distributor
     (b)   H-line coupler, with power distributor
     (c)   Ring coupler (ring hybrid)
     (d)   Ring feeder coupler (rat-race coupler)

In the case of the 3 dB coupler ($P_1/2 = P_2 = P_3$):

$$Z_{L1} = Z_L \text{ and } Z_{L2} = Z_L/\sqrt{2}. \tag{3.27}$$

in the case of the H-line coupler (Figure 3.20b):

$$P_2/P_3 = \frac{Z_L Z_{L3} + \frac{1}{2} Z_{L2}^2 \frac{Z_{L1}}{Z_L}(1 - (Z_L/Z_{L1})^2}{Z_{L1}Z_{L3} - Z_{L2}^2}$$

$$Z_{L1}^2 = Z_L^2 (2\frac{Z_{L1}Z_{L3}}{Z_{L2}^2} - 1) \tag{3.28}$$

in the case of power distribution $P_2 = P_3$, from eqn. 3.28:

$$Z_{L1}/Z_L = \sqrt{2} + 1 \text{ and } Z_L Z_{L3} = Z_{L2}^2 \sqrt{2} \tag{3.29}$$

With $Z_L = 50\,\Omega$ we accordingly obtain $Z_{L1} = 20.7\,\Omega$. If the value $Z_{L3} = 50\,\Omega$ is chosen for the dimensioning of an H-line coupler, we derive $Z_{L2} = 42.02\,\Omega$.

Figure 3.20d shows another type of coupler, known as the ring feeder coupler (or rat-race coupler). The electrical properties are similar to those of the couplers described previously, although the line lengths in this case are greater. This leads to couplers of this type having the behaviour of narrowband elements.

3.2.3.2   Line couplers

With stripline couplers, two lines running parallel to one another are located next to one another or, with stronger couplings, above one another. In this case, power is coupled from one power source to the other. Figure 3.21 shows a line coupler of this type. The lines are coupled to one another on the line length $l = \lambda/4$. It can be seen that lines which lie in a plane in accordance with Figure 3.21 can decouple power values which lie about 6 dB or more below the power P_0 fed in at port (1).

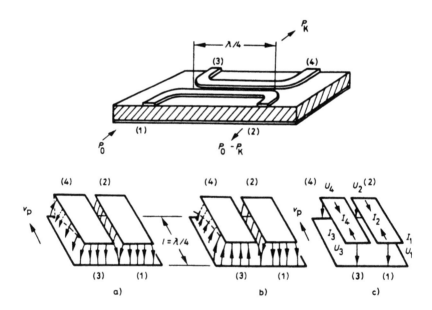

Figure 3.21   *Line directional couplers in microstrip technology; coupler link $l = \lambda/4$, P_0 = power being fed in (1), P_k = decoupled power (3)*

    (a)   Common-mode wave in coupled-line system, wave impedance Z_{0e}
    (b)   Push-pull wave in coupled line system, wave impedance Z_{0o}
    (c)   Voltages and currents at ports (1) to (4)

The coupling attenuation a_k provides the decoupling value:

$$a_k/dB = 10 \log P_1/P_3 = 10 \log P_0/P_k = 20 \log 1/k \qquad (3.30)$$

In this context, k is designated as the coupling factor.

$$k = (Z_{oe} - Z_{oo})/(Z_{oe} + Z_{oo}) \qquad (3.31)$$

The directional coupler attenuation (the directivity) a_D is given as a dimension for the decoupling factor. This amounts to

$$a_D/dB = 10 \log P_3/P_4 \qquad (3.32)$$

In ideal circumstances, at midfrequency, $P_4 = 0$ (i.e. $a_D \rightarrow \infty$).

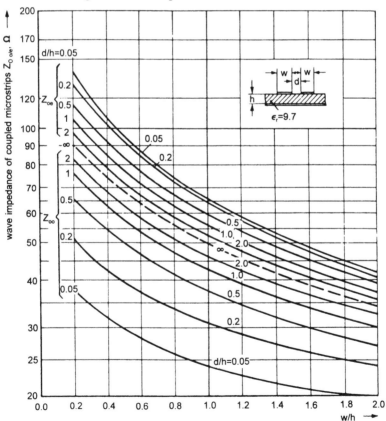

Figure 3.22   *Wave impedance of common-mode and push-pull wave Z_{oe}, Z_{oo}*

In the case of couplers shown in Figure 3.21, coupling values are usually measured in the order of 6 to 20 dB (with $a_D \approx 20$ dB).

For line couplers with coupling coefficients larger than $k = 0.5$ ($a_k = 6$ dB), the interval between the two striplines is so narrow that there is a risk of mutual contact. Accordingly, solutions are sought for the range $k = 0.5$ which will guarantee a narrower coupling for the lines.

Example: In accordance with Figure 3.21, with matching (Z_{in} (at port (1)) $= Z_L = 50\ \Omega$), it is intended to feed a power of $P_0 = 10$ mW into a coupled stripline. Given: substrate $\varepsilon_r = 9.7$; $h = 0.635$ mm, $f = 6$ GHz, $s = 0.635$ mm. How large is the power at the ports (2), (3) and (4), with matching, for the ideal attenuation-free situation? What is the length of the line coupling?

Solution:
(1) $\lambda_0 = 5$ cm, $\lambda_\varepsilon = F_v \lambda_0 = 19.5$ mm, with $w/h = 1$ (50 Ω), Figures 3.3 and 3.4.
(2) Length l of line coupling: $l = \lambda_\varepsilon/4 = 4.88$ mm
(3) Wave impedance of common-mode wave: $Z_{oe} = 56.5\ \Omega$, Figure 3.22
    Wave impedance of push-pull wave: $Z_{oe} = 42.5\ \Omega$
(4) Coupling coefficient $k = 14/99 = 0.14$ as per eqn. 3.31
(5) $P_k = P_{(3)} = P_0 k^2 = 0.2$ mW; $a_k = 17$ dB
(6) $P_{(4)} = 0$ in the ideal case: $P_{(2)} = P_0 - P_k = 9.8$ mW

3.2.3.3 PIN diode switches
PIN diodes are capable of switching relatively high microwave powers, and for this reason this type of diode is preferred over other diodes for switching processes.

Figure 3.23 shows a typical PIN diode switch in stripline technology. In the frequency range in question, the diodes have the effect of short-circuit elements, while in the nonoperating range the diodes connected in parallel to the line source have no effect on the transmitted power because of their high resistance.

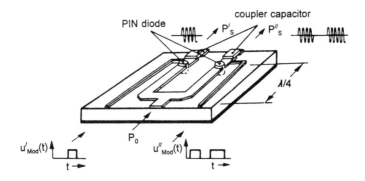

Figure 3.23 *Pulse modulator, PIN diode push-pull switch*

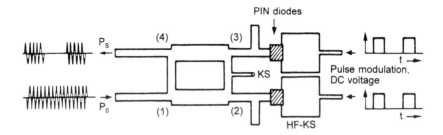

Figure 3.24   *PIN diode hybrid switch*

(1)-(4)   inputs of square hybrid; P_0 = oscillator power; P_s = pulse-modulated
output power; KS = short-circuit

A diode which is short-circuited by the modulation voltage has the effect of a no-load
resistor, owing to the $\lambda/4$ transformation of the line, for example at the front-left line
branch point. The power P_0 fed in is then conducted free of reflection to the upper
left-hand output (P_s'), if the right-hand diode is short-circuited and the left-hand
diode is unloaded, i.e. high-resistance. Figure 3.24 shows a PIN diode hybrid switch.

3.2.3.4 Microwave mixers

Figure 3.25 shows a matching contact mixer in stripline technology. The diodes
connected in push-pull format have the effect, with the power distribution arrangement made here, of providing an overlay in correct voltage relationship of the wave
components of the intermediate frequency signal. The IF voltage is short-circuited

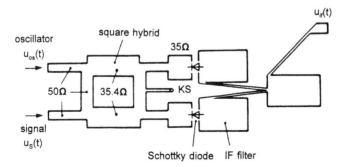

Figure 3.25   *Microwave mixer, push-pull switch with Schottky diode, IF filter in
π-circuit according to Figure 3.18c*

over a spur line in the high-frequency section, of λ/4 in length, so that this voltage can be conducted via the IF filter to the carrier frequency output, and is decoupled from the high-frequency inputs.

3.2.3.5 Transmission converter

Figure 3.26 shows a transmission converter at 6.7 GHz in microstrip technology. Two reception signals, in diversity reception mode, come from two different antenna lines. These two signals are brought into the intermediate frequency setting jointly with the oscillator signal. The two IF signals overlay one another in this situation in correct phase relation. By adding the reception voltages, shrinkage phenomena can be avoided.

Figure 3.26   *Transmission converter for DRS 140 radio relay system, at 6.7 GHz*

> Microstrip technology: Al_2O_3 ceramic substrate (converter); ferrite substrate (circulator)

3.2.4 Transistor amplifier in stripline technology

Because of their design with one side open, microstrip is well-suited for use in transistor circuits. Matching circuits for transistors can be represented easily, for example, by the series connection of lines of different wave impedances or by unloaded spur lines. Bipolar transistors (npn transistors) are suitable for use as amplifier elements up to about 6 GHz and are used both for low-noise amplification as well as for power amplification. At frequencies above 6 GHz and up to 25 GHz, field effect transistors are preferred as low-noise amplifier elements and as power amplifiers.

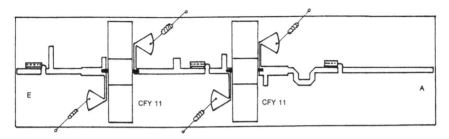

Figure 3.27   *Low-noise 10 GHz amplifier layout*

Substrate: Epsilan 10; G = 12 dB; F = 3.5 dB [20]

Figure 3.27 shows the layout of a two-stage field effect transistor amplifier (CFY 11). This amplifier operates in the frequency range from 10 to 10.5 GHz. Total amplification attained in this situation is 12 dB. Noise factor of the 10 GHz amplifier is about 3.5 dB.

For DC voltage unblocking, elements referred to as DC blocks are used between the inputs and outputs of the transistors. These are λ/4 long, coaxial cable sections unloaded at the end, which simulate an HF short-circuit at the input of the line section and which simultaneously cause isolation of DC voltages. The line arrangements, in the form of a circle segment, are no-load stripline formations, λ/4 long, which transform a short-circuit to a high resistance λ/4 spur line. The end of the spur line, directly at the transistor input, is high resistance, and therefore does not impose a load on the line run. DC voltage inputs (e.g. via ferrite reactance coils) can be soldered to the line segment without affecting the HF section.

In the high-frequency sector, in certain cases there is a shift today towards connecting interface circuits permanently to the transistor housing. In the K-band frequency range (18 to 26.5 GHz), transistors are available on the market which already contain the interface circuit in microstrip technology. Figure 3.28 shows transistors of this type.

Figure 3.28   *Integrated transistor circuit in microwave range (K-band)*

Chapter 4

Amplifier transistor techniques

P. Pauli

4.1 Introduction

The development of semiconductor components, in particular of transistors, for use at frequencies in the GHz range has very recently progressed at a pace hitherto undreamt of. Three points in particular have contributed to this:

- new discoveries and improved mastery of the process of crystal manufacture and doping in the context of semiconductors from GaAs, InP, and other mixed crystals

- the techniques which have also forged ahead in the interim with regard to the development, manufacture, and investigation of semiconductor components, such as cutting and balancing processes with laser beams, or the investigation of the product obtained under a scanning electron microscope

- new and more informative models that describe the HF behaviour of the components, and new, often computer-controlled measuring devices (in the frequency range up to 110 GHz), with which the characteristic values and parameters of the transistors can be determined relatively quickly and simply over the entire frequency range of interest.

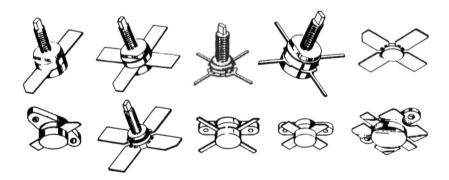

Figure 4.1   *HF transistors*

In most cases, users are not so interested in how HF transistors are manufactured, and what with. Much more important are the measurement data referred to in the last point to be able to determine optimum dimensioning of the circuit arrangement. In addition to this, users should know about ways of examining the quality of a circuit and how it might be improved. The following pages provide a number of possible solutions to these questions.

4.2 Possible ways of describing transistor properties

From the field of general electrical engineering, a transistor can also be regarded as a quadripole (or 2-port). This means that its properties are to be described by one of the many parameters from quadripole theory. Depending on the wiring conditions of the quadripole (no load, short-circuit, matching) a number of parameter systems can be eliminated as far as higher frequencies are concerned, while others, in particular the S-parameters, offer particular advantages. First, however, consider briefly the significance of the Y- (admittance value) parameters, because these parameters are still encountered very frequently in transistor data sheets to describe the properties of transistors.

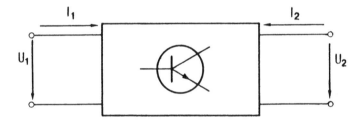

Figure 4.2   *Quadripole with reference arrow system*

4.2.1 Admittance Y-parameters

The Y-parameters are derived from the quadripole equivalent circuit diagram shown in Figure 4.3.

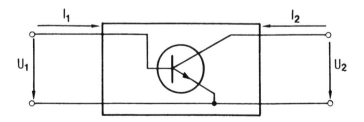

Figure 4.3   *Transistor in common emitter mode*

The following complex formulation may be used for the special case of sinusoidal small-signal values of current and voltage:

$$J_1 = Y_{11} \cdot U_1 + Y_{12} \cdot U_2$$

$$J_2 = Y_{21} \cdot U_1 + Y_{22} \cdot U_2$$

To measure the individual parameters, unfavourable short-circuit measurements must be taken for the quadripole. From these, we derive the Y-parameters:

$$Y_{11} = g_{11} + jb_{11} = \left. \frac{J_1}{U_1} \right|_{U_2 = 0} \qquad \text{input (short–circuit)} \qquad \text{e.g.}$$
$$\text{conductivity value} \qquad \text{mS}$$

$$Y_{12} = g_{12} + jb_{12} = \left. \frac{J_1}{U_2} \right|_{U_1 = 0} \qquad \text{reverse transfer} \qquad \text{e.g.}$$
$$\text{admittance} \qquad \text{mA/V}$$

$$Y_{21} = g_{21} + jb_{21} = \left. \frac{J_2}{U_1} \right|_{U_2 = 0} \qquad \text{forward transfer} \qquad \text{e.g.}$$
$$\text{admittance} \qquad \text{mA/V}$$

$$Y_{22} = g_{22} + jb_{22} = \left. \frac{J_2}{U_2} \right|_{U_1 = 0} \qquad \text{output (short–circuit)} \qquad \text{e.g.}$$
$$\text{conductivity value} \qquad \text{mS}$$

The data book values for the Y-parameters, as shown in Figure 4.4, for example, can be read off and divided into real and imaginary parts.

As shown in Figure 4.4a, at its working point, e.g. at $C_{CB} = 10$ V and $J_C = 2$ mA, and at a frequency $f = 100$ MHz, the transistor has an input admittance value of

$$Y_{11} = g_{11} + jb_{11} = 45 - j\,28 \text{ mS}.$$

The important point is, as already mentioned, that these admittance parameters are only exactly applicable to small amplitudes, and are only determined for open or short-circuit circumstances.

Accordingly, the effect of the actual wiring of the transistor quadripole in terms of impedance, and its calculability, are considerably more indicative. In this case, one may reckon on the Y-parameters given in the data sheets for the first overview in the prescribed frequency range, and, taking into account the input circuit (with the source conductivity value, subscript Q) and the output load (load conductivity value, subscript L), the following new parameters are derived:

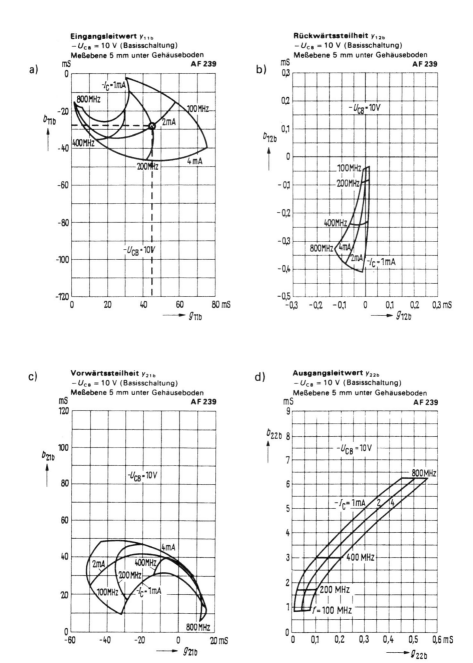

Figure 4.4   *Y-parameters*

−U$_{CB}$ = 10V (grounded base circuit)

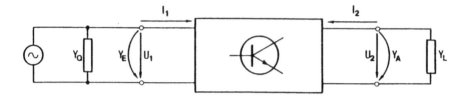

Figure 4.5   *Connected transistor quadripole*

input conductivity value
(without source circuit)

$$Y_E = \frac{J_1}{U_1} = Y_{11} - \frac{Y_{12} \cdot Y_{21}}{Y_{22} + Y_L} \approx Y_{11}$$

output conductivity value
(without taking account of
the load conductivity value)

$$Y_A = \frac{J_2}{U_2} = Y_{22} - \frac{Y_{12} \cdot Y_{21}}{Y_{11} + Y_Q} \approx Y_{22}$$

voltage amplification

$$Y_U = \frac{U_2}{U_1} = \frac{-Y_{21}}{Y_{22} + Y_L} \approx \frac{-Y_{21}}{Y_L}$$

current amplification

$$Y_J = \frac{J_2}{J_1} = \frac{Y_{21} \cdot Y_L}{Y_{11} \cdot (Y_{22} + Y_L) - Y_{12} \cdot Y_{21}} \approx \frac{Y_{21}}{Y_{11}}$$

These equations apply to all three types of basic transistor circuits, albeit with the parameters which belong in each case for the particular kind of circuit required (grounded-emitter [E], grounded-base [B], or grounded-collector [C] circuit).

4.2.2 Scattering or S-parameters

To describe the transistor as an amplifier and to plan the dimensions for the input and output circuit, the S-parameters are the best to use at high frequencies. As explained in the previous Section, to determine the S-parameters, the measured object is embedded in a line system with a quite specific wave impedance. The practical user must be aware, in this situation, that this characteristic wave impedance Z_W, Z_L, Z_0, or for hollow conductors Z_F or Z_g, can be specified in the various different data books or application specifications.

In measurement engineering terms, the S-parameters have in the meantime become easy to determine with the availability of wideband sweep frequency generators with good linearity, wideband network analysers, vector voltmeters, directional couplers in every frequency range, and optimum terminating resistors.

HF transistors must be fitted into suitable measuring adapters (see Figure 4.6) for the measurement of S-parameters, in which the current supply and the HF input, as discontinuity-free as possible, are in most cases provided over striplines.

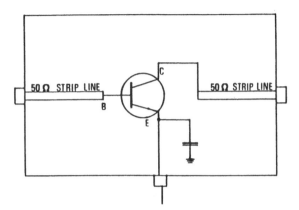

Figure 4.6   *Transistor adapter for S-parameter measurement*

S-parameters can be measured according to magnitude and phase as complex quantities with vector voltmeters or network analysers. To do this the measurement adapter and the transistor must be connected to directional couplers at the input and output (as in Figure 4.7) and lines from the wave impedance Z_L.

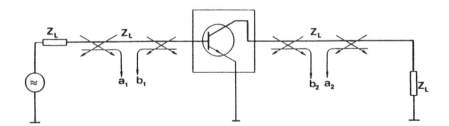

Figure 4.7   *Measuring arrangement*

With this measuring arrangement the fractions of the wave amplitudes a_1 entering the quadripole, and reflected fractions b_1 at the input, can be determined. Figure 4.8 again shows, in schematic form, the designation of the individual wave magnitudes and the arrangement of the S-parameter designations.

If the transistor adapter is reversed, in other words with the transistor now being actuated from the output, and its response measured at the input, one can derive the values for b_2 and a_2 and therefore the output S-parameter (reflection coefficient).

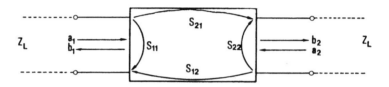

Figure 4.8   *Quadripole with S-parameters*

The transistor can be described just as accurately with the quadripole equations in S-parameter form as it can with the other parameter systems, with the advantage that the values with matching at the input and output now correspond much more closely to the real operational situation.

$$b_1 = S_{11}\,a_1 + S_{12}\,a_2$$

$$b_2 = S_{21}\,a_1 + S_{22}\,a_2$$

These S-parameters, which are capable of being easily measured, in practice have very readily appreciable significance:

$S_{11} = b_1/a_1$ is the input S-parameter
$S_{22} = b_2/a_2$ is the complex output S-parameter
$S_{21} = b_2/a_1$ describes the forward transfer coefficient behaviour
$S_{12} = b_1/a_2$ is the reverse transfer coefficient behaviour of the transistor.

The user need only know how to obtain the S-parameters which are sought from the very wide variety of representations in the individual data books. Figures 4.9 to 4.12 are intended to show a number of possibilities of the form in which S-parameter data relating to transistors may be presented; the same required data can, however, be obtained from each Figure.

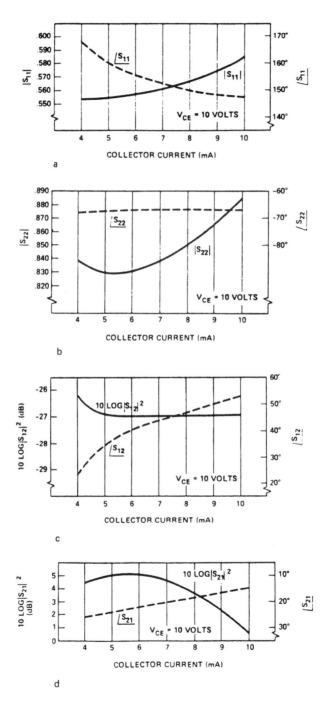

Figure 4.9   *Parameter data in magnitude and phase form*

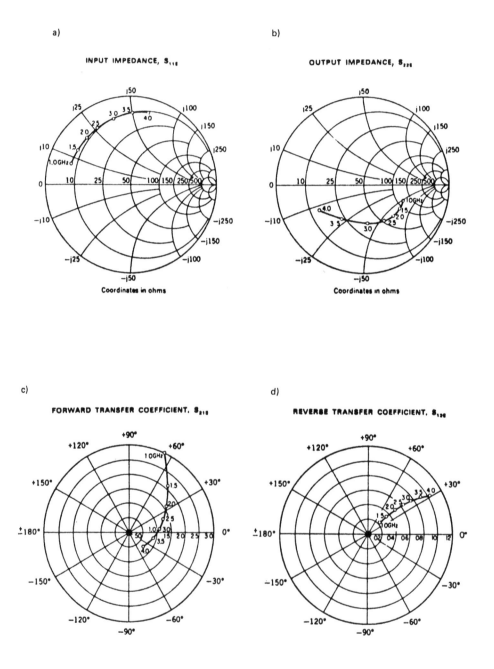

Figure 4.10   *S-parameter data in Smith diagram and polar co-ordinates*

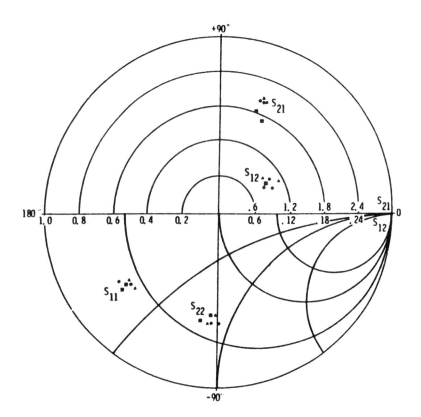

Figure 4.11   *S-parameters in combination diagram*

Common source S Parameters, Z_o = 50 Ω

Operating point: V_{DS} = 4 V, I_D = 10 mA

Frequency MHz	S 11 Mag.	Ang.	S 21 Mag.	Ang.	S 12 Mag.	Ang.	S 22 Mag.	Ang.
2000	0.90	− 44	2.28	132	0.06	57	0.74	− 28
2500	0.85	− 56	2.24	119	0.07	48	0.70	− 36
3000	0.79	− 66	2.17	108	0.08	42	0.67	− 42
3500	0.74	− 79	2.18	97	0.09	33	0.63	− 49
4000	0.69	− 91	2.16	86	0.09	28	0.59	− 57
4500	0.65	−105	2.08	74	0.10	22	0.57	− 65
5000	0.60	−119	2.11	63	0.11	16	0.51	− 72
5500	0.56	−135	2.04	50	0.10	6	0.48	− 83
6000	0.50	−155	2.04	38	0.10	2	0.42	− 91
6500	0.49	−172	1.89	25	0.10	− 6	0.39	−106
7000	0.46	172	1.87	14	0.10	− 8	0.36	−115
7500	0.48	150	1.77	1	0.09	−14	0.31	−133
8000	0.50	134	1.65	− 12	0.09	−19	0.31	−156
8500	0.51	126	1.53	− 20	0.08	−18	0.34	−164
9000	0.54	111	1.42	− 34	0.08	−23	0.33	173
9500	0.54	110	1.31	− 40	0.08	−19	0.38	168
10000	0.57	93	1.24	− 55	0.08	−26	0.36	147
10500	0.57	97	1.13	− 58	0.08	−22	0.43	147
11000	0.56	80	1.14	− 72	0.10	−29	0.38	136
11500	0.58	78	1.06	− 78	0.10	−32	0.45	123
12000	0.56	60	1.09	− 91	0.11	−39	0.40	118

Operating point: V_{DS} = 4 V, I_D = 30 mA

Frequency MHz	S 11 Mag.	Ang.	S 21 Mag.	Ang.	S 12 Mag.	Ang.	S 22 Mag.	Ang.
2000	0.84	− 55	3.40	124	0.05	55	0.69	− 29
2500	0.76	− 69	3.23	110	0.06	46	0.65	− 36
3000	0.68	− 81	3.03	98	0.06	42	0.61	− 41
3500	0.62	− 96	2.92	86	0.07	34	0.57	− 48
4000	0.56	−110	2.79	76	0.07	34	0.54	− 55
4500	0.52	−124	2.64	65	0.07	29	0.52	− 62
5000	0.48	−142	2.61	53	0.08	25	0.46	− 67
5500	0.45	−159	2.46	41	0.08	19	0.44	− 77
6000	0.42	177	2.40	29	0.08	19	0.38	− 83
6500	0.43	162	2.21	18	0.08	13	0.35	− 99
7000	0.42	146	2.14	7	0.08	13	0.34	−106
7500	0.46	128	2.01	− 4	0.08	10	0.29	−125
8000	0.51	114	1.87	− 16	0.09	6	0.28	−148
8500	0.51	109	1.74	− 24	0.09	5	0.31	−158
9000	0.55	96	1.61	− 38	0.09	−	0.30	178
9500	0.54	95	1.49	− 43	0.10	−	0.35	173
10000	0.58	80	1.41	− 58	0.10	− 8	0.33	151
10500	0.57	84	1.30	− 60	0.11	− 8	0.41	151
11000	0.57	67	1.30	− 75	0.12	−17	0.35	140
11500	0.58	66	1.22	− 81	0.12	−22	0.43	126
12000	0.57	46	1.23	− 95	0.14	−32	0.38	123

Figure 4.12   *S-parameters in tabular form*

4.3 Important parameters in Smith chart

Depending on which frequency range one is working in at the time, which data book happens to be to hand, and what task one is seeking to solve with the dimensioning of a transistor amplification circuit, different information about the transistor may be required. One useful aid to the high-frequency engineer in representing complex impedances is known as the Smith chart. The grid can, of course, also be used as a complex conductance plane, as Figures 4.13 and 4.14 show. In this case, the original loci r = const. in the Y-plane now function as loci of constant conductance, and the loci x = const. now function as loci of constant susceptance.

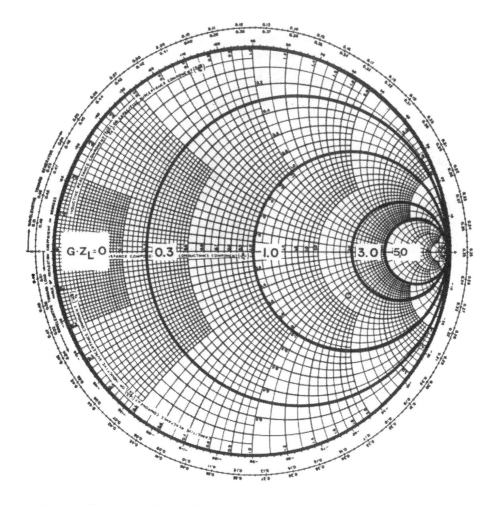

Figure 4.13   *Smith chart with lines of constant conductance*

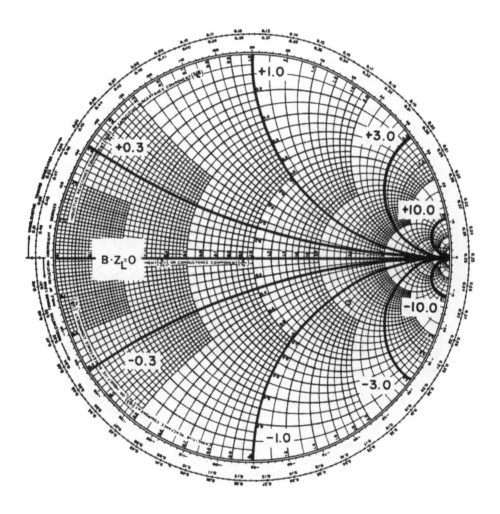

Figure 4.14   *Smith chart with lines of constant susceptance*

4.3.1 Transistor impedances derived from Y-parameters

If the Y-parameters are known from data book details, one can obtain input and output conductance values for the transistor by entering the real components and reactances in the complex admittance value plane.

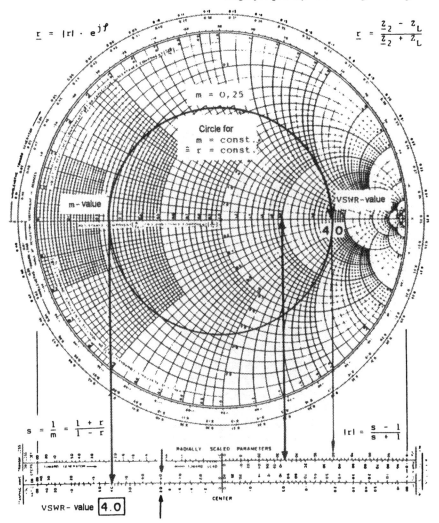

$$\underline{r} = |r| \cdot e^{j\varphi}$$

$$\underline{r} = \frac{\underline{z}_2 - \underline{z}_L}{\underline{z}_2 + \underline{z}_L}$$

$$s = \frac{1}{m} = \frac{1 + r}{1 - r}$$

$$|r| = \frac{s - 1}{s + 1}$$

Figure 4.15 *Connection between m, VSWR, and |r|*

4.3.2 Transistor impedances derived from S-parameters

S_{11} and S_{22} are complex reflection coefficients which are incurred by the fact that the input and output impedance of the transistor is not matched to the wave impedance Z_L of the connection line.

It is therefore possible to allocate to each complex reflection coefficient a quite specific complex impedance which is causing the reflection, and vice versa. This also enables one, by suitable designation or by the appropriate scales, to enter the complex reflection coefficient in an impedance Smith diagram.

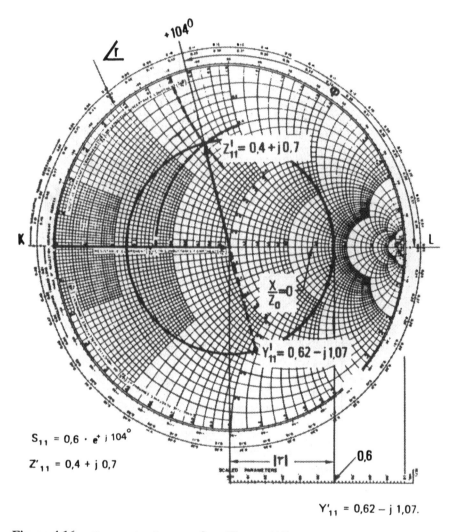

Figure 4.16 *Connection between* S_{11}, Z'_{11} *and* Y'_{11}

The connection between lines of constant $|r|$ and the known m loci can be calculated:

$$|r| = \frac{1-m}{1+m} = \frac{s-1}{s+1}$$

s = VSWR = voltage standing wave ratio, m = matching factor. An extra angle scale must be provided on the circumference of the Smith diagram for the angle of the reflection coefficient. With these details, it can be seen from the Smith diagram that an input reflection coefficient

$$S_{11} = 0.6 \cdot e^{+j\,104\,°}$$

was caused by a standardised complex input impedance of the transistor of $Z'_{11} = 0.4 + j\,0.7$. This corresponds to a standardised input conductance of $Y'_{11} = 0.62 - j\,1.07$.

4.4 Neutralisation

The parameters Y_{12} or S_{12} describe the reciprocal effect of the output side on the input (control) side of the transistor quadripole. Particularly with amplifiers with high output power, even a slight perturbation can cause such power feedback on the control side of the transistor that, with the matching phase angle, an unwanted tendency to oscillation or even self excitation may occur. To prevent this effect, we must try to neutralise this internal feedback in the quadripole by means of an external circuit.

If the undesired signal that is being caused by the internal feedback is known (caused in most cases by capacitative and ohmic couplings in the semiconductor), one must create a neutralisation circuit in such a way that a signal is injected into the control input which is of the same value in terms of amplitude, but opposed in terms of phase. The phase opposition is obtained by the fact that the signal that is to be fed back is drawn from a transformer with windings in the counter direction.

The dimensioning of the neutralisation circuit can be carried out once we are aware of the values Y_{12} or S_{12}.

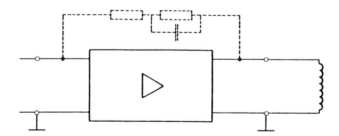

Figure 4.17   *Internal feedback (dashed components)*

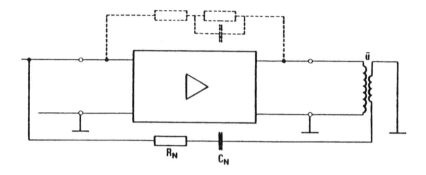

Figure 4.18 *Wiring of quadripole for neutralisation*

Example: If we know, for example, $Y_{12} = -0.08 - j0.4$ mS and assume a transformer with an impedance transformation ratio $ü = 1 : 10$, the neutralisation conductance value required can be calculated:

$$Y_{Neutr.} = -\frac{Y_{12}}{ü} = \frac{0.08 \text{ mS} + j0.4 \text{ mS}}{0.1} = 0.8 \text{ mS} + j\,4 \text{ mS}$$

$$Y_N = G_N + jB_N = \frac{1}{1.25 \text{ k}\Omega} + j\omega\ 4.6 \text{ pF}$$

Conversion into a series circuit gives

$$R_S = \frac{R_p}{1 + (\omega C_p\,R_p)^2} = 48\ \Omega = R_N$$

$$C_S = \frac{1 + (\omega C_p\,R_p)^2}{C_p\omega^2 \cdot R_p^2} = 6.7 \text{ pF} = C_N.$$

In most cases, the capacitative fraction is so overwhelming that one can omit R_N altogether. With transistor power amplifiers, exact neutralisation is often difficult because of the feedback capacitance being dependent on the modulation. On the other hand, transistors with so little feedback have now become available for high frequencies that a neutralisation circuit is often superfluous.

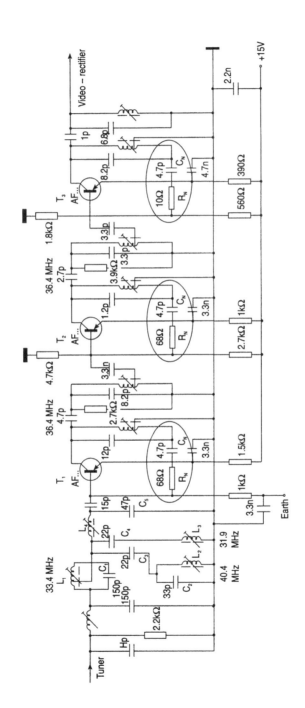

Figure 4.19 *Example of circuit for inclusion of neutralisation components in TV-IF amplifier stage*

4.5 Matching

4.5.1 Line matching

The Sections relating to Y- and S-parameters showed how the current and voltage amplification of the transistor depends on its connection with source and load impedance. In the S-parameter system one can see that, with nonmatched inputs and outputs, input and output reflections arise, and therefore in this case too the degree of efficiency of the amplifier circuit drops.

Now consider which criteria a load component must fulfil which is connected to the complex output impedance of an active two-port network (approximately equivalent to the output side of the transistor) for maximum active power to be transferred to it.

In Figure 4.20:

$$\underline{I}_2 = \frac{\underline{U}_0}{\underline{Z}_i + \underline{Z}_2} \; ; \; \underline{U}_2 = \underline{Z}_2 \cdot \underline{I}_2 \; ; \; Z_i = R_i + jX_i; \; \; Z_2 = R_2 + jX_2$$

The active power which passes into Z_2 is calculated from the current and the real part $\mathrm{Re}(Z_2)$:

$$P_2 = |\underline{I}_2|^2 \cdot R_2 = \frac{\underline{U}_0{}^2}{|\underline{Z}_i + \underline{Z}_2|^2} \cdot R_2 = \frac{\underline{U}_0{}^2}{(R_i + R_2)^2 + (X_i + X_2)^2}$$

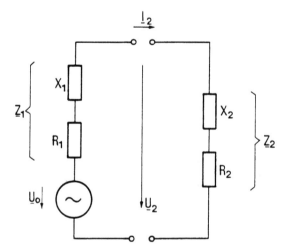

Figure 4.20   *Quadripole as voltage source U_0 with internal impedance*
$Z_i = R_i + jX_i$

On the assumption that U_0 and Z_i are fixed preset values, one can attempt to optimise Z_2 in such a way that the maximum active power passes into the load component. To do this, the expression for P_2 is differentiated according to the possibly variable R_2 and X_2, and the derivatives are set at the equivalent of zero. From this we derive the conditions: $R_2 = R_i$ and $X_2 = -X_i$; in other words, the requirement that, for maximum power output to a load component Z_2, the component must be conjugately matched in a complex manner to the amplifier output.

4.5.2 Dimensioning matching circuits for HF transistors

For practical amplifier circuits, the additional task frequently arises of transferring the power to the transistor and the amplified power to the load via transmission lines of standard wave impedance.

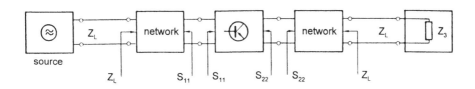

Figure 4.21   *Wiring of transistor quadripole in practice*

It is known that lines (with wave impedance Z_L) can be connected reflection-free and independent of frequency only to a real impedance value of magnitude Z_L. Accordingly, the task arises for almost every transistor installation in an amplifier circuit of matching the transistor via a transformation network to a real line impedance (standard values for Z_L are, for example, 50, 60, or 75 Ω).

For further explanations of the dimensioning, it should be assumed that the S-parameters of the transistor are available.

4.5.3 Noise matching and amplification optimisation

It is known that a transistor, when connected to a generator with a specific internal impedance Z_{QF}, produces a minimal noise; this is referred to as noise matching. It is also known that, with a quite specific generator internal impedance Z_{QV}, a maximum amplification is possible, and, at another resistance value of the size Z_{QP}, the maximum power transfer is derived. Unfortunately, these three impedance values are in practice not identical. The result of this is that when planning the dimensions we must in most cases reach a compromise between minimal noise and maximum amplification. The use of the S-parameters makes reaching such a compromise substantially easier.

One can enter the dependency of the amplification and the attainable noise figure of a transistor with source output impedance S^*_{11} in the complex reflection coefficient plane. In this situation one finds that there is an optimum value S^*_{11} as the source output impedance, at which the minimal noise figure F is attained. All other impedance values produce higher noise values; the corresponding values of the noise figure obtained can be read off the curve (solid circle loci in Figure 4.22). For input stages where a particularly high signal-to-noise ratio obtains, the transformation network is designed from the point of view of optimum noise matching.

In Figure 4.22 one can identify from the curve segment (broken circle loci), which represent the dependency of the attainable amplification, that in this case about 4 dB is sacrificed in amplification.

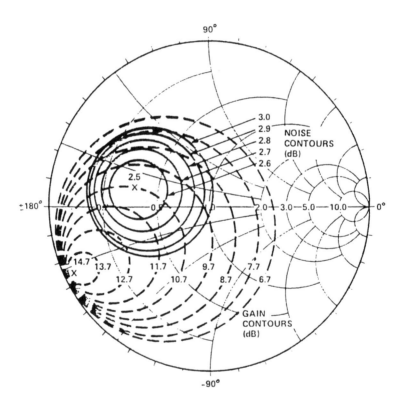

Figure 4.22   *Loci of constant noise and constant amplification*

In Figure 4.23, a number of outstanding points are identified in the complex reflection coefficient plane. In this case the values designated here as Γ already represent the values S^*_{11} and S^*_{22}, which should be offered to the transistor from the circuit.

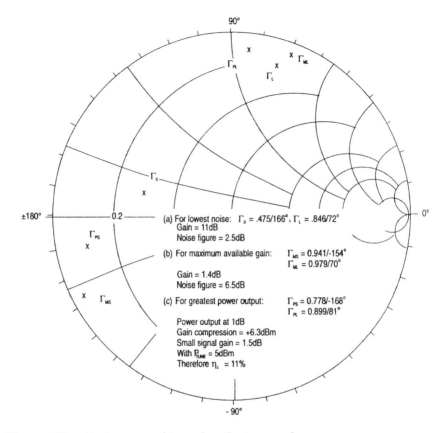

The following values appear within the Smith diagram:

90°

X

Γ_R X Γ_{ML}

X

Γ_L

Γ_0

X

±180°   0.2   0°

Γ_{PS}

X

X Γ_{MS}

(a) For lowest noise: $\Gamma_0 = .475/166°$, $\Gamma_L = .846/72°$
Gain = 11dB
Noise figure = 2.5dB

(b) For maximum available gain: $\Gamma_{MS} = 0.941/-154°$
$\Gamma_{ML} = 0.979/70°$

Gain = 1.4dB
Noise figure = 6.5dB

(c) For greatest power output: $\Gamma_{PS} = 0.778/-168°$
$\Gamma_R = 0.899/81°$

Power output at 1dB
Gain compression = +6.3dBm
Small signal gain = 1.5dB
With $P_{TUNE} = 5dBm$
Therefore $\eta_c = 11\%$

- 90°

Figure 4.23   *Optimum matching values for input and output*

4.5.4 Developing the matching circuit

On the output side the dimensions are usually arranged in accordance with the rules of power matching. In this case the following steps are required.

(a) S_{22} must be known (for the exact solution, S_{22} must be calculated for the source-side wiring of the transistor, from the point of view of noise matching).

(b) S_{22} is entered in the Smith diagram.

(c) S^*_{22}, the complex conjugate of S_{22}, is calculated.

(d) The complex impedance value Z^*_{22}, which belongs to S_{22}, is determined.

(e) An impedance transformation is then carried out passing from Z^*_{22} to the matching point 1, which corresponds to the value of the wave impedance of the connection line (if nothing else is specified, nowadays a wave impedance of $Z_L = 50\ \Omega$ is taken as a precondition).

The transformation path is to be selected in such a way that line lengths which may be required for the transistor connection, for current inputs, or for matching possibilities are taken into account. In addition, it is important to choose the right direction of rotation in line transformation in the Smith diagram for the transformation from the transistor to the consuming component (namely, anticlockwise).

Example: Enter $S*_{22} = 0.627\ e^{+j\ 106°}$
against $S_{22} = 0.627\ e^{-j\ 106°}$

in the Smith diagram, and read off from this:

$$z^*_{22} = 0.32 + j0.7 \Rightarrow \quad y^*_{22} = 0.55 - j1.15$$

$$Z^*_{22} = 17.5 + j35\ \Omega \Rightarrow \ Y^*_{22} = 0.012 - j0.023\ S$$

The conductance value Y is derived by inversion of the resistance Z. For this example, the inductive susceptance for $Y*_{22}$ is represented by a parallel-connected *spur line*. The value of the inductive susceptance should amount to $jb = -j1.15$

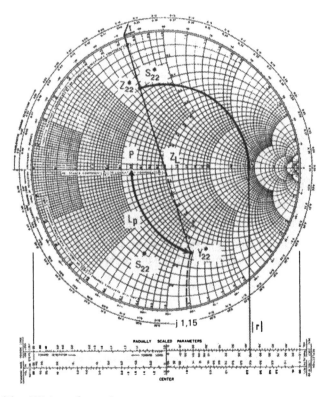

Figure 4.24   *Wiring of transistor output*

(normalised) or jB = – j0.023 S. For the input susceptance of a spur line which is short-circuited at the end,

$$jB = -j\frac{1}{Z_L} \cdot \cot\frac{2\pi l}{\lambda}$$

For a line which is open-ended,

$$jB = -j\frac{1}{Z_L} \cdot \tan\frac{2\pi l}{\lambda}$$

Figure 4.25 shows in graphical terms the reactance curve of a spur line short-circuited at the end, and Figure 4.26 that of an unloaded spur line.

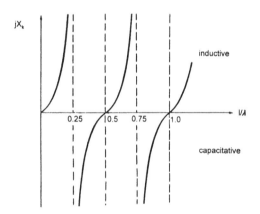

Figure 4.25   *Reactance curve with short circuit (jX_k)*

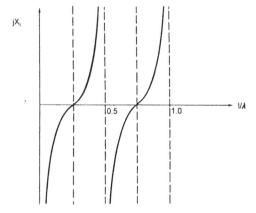

Figure 4.26   *Reactance curve of unloaded line (jX_l)*

One can see here, for example, a line which is short-circuited at the end showing an inductive behaviour $l < \lambda/4$, and capacitative behaviour with $\lambda/4 < l < \lambda/2$. In practice, the length of the spur lines is frequently chosen to be exactly $\lambda/8$ or $3\lambda/8$, since for these the tangent function assumes the value 1. For inductive reactance, the spur line which is short-circuited at the end must be $\lambda/8$ long, and the open line $3\lambda/8$ long. The size of the reactance value required is now dependent only on the wave impedance of the spur line section, in other words on its conductor width.

For the wave impedance of the spur line Z_{LS}:

$$Z_{LS} = \frac{1}{|B|} = \frac{1}{0.023} = 43.5 \ \Omega$$

By adding this spur line, we obtain in the Smith diagram (Figure 4.24) the now real and standardised conductance value $g = 0.6$ at the point P, corresponding to a resistance of 83 Ω. Over a line $\lambda/4$ long, this real value of 83 Ω can be matched to the wave impedance of the input line of 50 Ω. the wave impedance of this $\lambda/4$ transformation section can be calculated to

$$Z_{Lt} = \sqrt{R_1 \cdot R_2} = \sqrt{83 \cdot 50} = 64.5 \ \Omega$$

Figure 4.27 shows the schematic composition of the output circuit of the transistor, which accordingly contains a conjugated-complex termination for maximum power transfer.

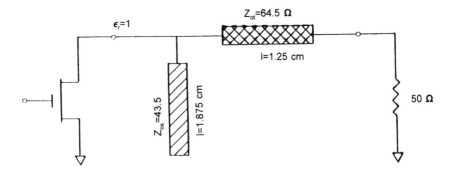

Figure 4.27   *Wiring layout of HF transistor for f = 6 GHz*

The layout in stripline technology is shown in Figure 4.28. In this context, the individual line sections are hatched in the same way on the output side in comparison with Figure 4.27. The lines marked in black are already provided with the wave impedance of 50 Ω.

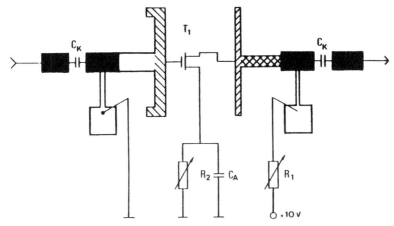

Figure 4.28 *Layout for transistor circuit in stripline technology*

Note here that the spur lines wired in parallel are not asymmetrical in respect of one side, but are allocated to individual conductance values which in each case are half as great, right and left of the longitudinal axis of the circuit.

The bandwidth of this amplifier is shown in Figure 4.29, and, despite the wavelength-dependent components in the passband range, still demonstrates usable linear behaviour.

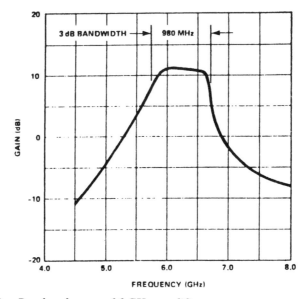

Figure 4.29 *Passband curve of 6 GHz amplifier*

For even greater wideband behaviour, the rules of wideband matching are to be applied for the matching transformation stages. This means that the matching circuit is to be resolved in several dual impedance sections, as takes place at the input of the circuit in Figure 4.30.

A precondition for succeeding with a wideband matching process, however, is the requirement that no further parasitic reactive component effects play a part in the matching component elements under consideration. This will only be assured if one can design the reactive components to be as small as possible; e.g. the longitudinal inductances in the form of direct bonded wires (see Figure 4.31) and the case capacitances are, if possible, already integrated into the semiconductor material as small islands, with the appropriate doping. In practice, transistors with internal matching of this type, as shown in Figure 4.32, can be purchased as IMPACs (input matched packages) or as AMPACs (all matched packages).

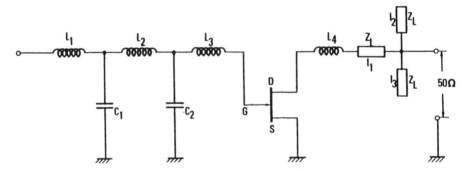

Figure 4.30 *Wideband matching of transistor input by means of concentrated components*

Figure 4.31 *HF transistor with wideband matching circuit*

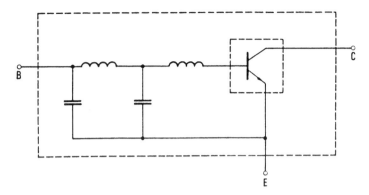

Figure 4.32   *Circuit diagram of IMPAC transistor*

Figure 4.33 shows the dependency and drop in amplification of an AMPAC transistor as a function of the size of the load impedance. Even if the optimum impedance partner Z_{opt} is supposed to possess a slightly capacitative component, a 50 Ω line can be connected to this component without further ado and without any additional matching effort, losing only about 0.25 dB in the amplification.

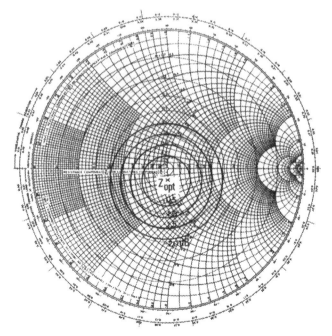

Figure 4.33   *Decrease in amplification of AMPAC as a function of load impedance*

Chapter 5

FET and bipolar power transistor amplifiers for 3–1000 MHz

K. Hupfer

5.1 Introduction

Ever since its invention by Shockley and Bardeen in 1947, the transistor has acquired a scope of application in the field of electronics as an amplifying component which would originally have been undreamt of. In addition to the large number of small-signal circuits, this semiconductor element has also established itself in the sectors of high-frequency power devices such as transmitters, industrial generators, medical microwave equipment, and radar systems. While transmitter amplifiers up to 10 kW were, until a few years ago, still the exclusive domain of tubes, there has now been an enormous and revolutionary swing in favour of HF power transistors. True, there is as yet still no semiconductor amplifier element which can produce this power; but, with the possibilities offered by modern circuit engineering, and in particular with the technology of wideband power distributors and combiners, it would be possible to connect together any desired number of individual transistors with a power rating, for example, of 100 W. In Germany, VHF transmitters have been developed with 10 kW output power which use, for example, about 20 transistors in the output stage. Japan has produced a 20 kW television transmitter with 100 transistors in the output stage; modern pulse transistors for radar applications produce up to 1 kW in a housing, and expensive travelling wave amplifiers are being replaced more and more by solid-state arrangements.

5.2 Composition of bipolar power transistors and FETs

5.2.1 The bipolar transistor

Although in the past few years the field-effect transistor has increasingly been found in high-frequency power amplifiers, the bipolar transistor is still the most widely-used amplifying semiconductor element. From short waves to the UHF range there are individual transistors available with an output power of some 200 W; and as twin (Gemini) components (two identical transistors in the same housing), there are

200 W amplifier elements available. Figure 5.1 shows a number of the housing forms used.

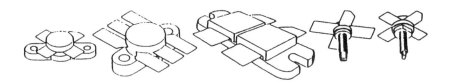

Figure 5.1   *Various different housings for HF power transistors*

The layout of a silicon power transistor is illustrated briefly with the aid of Figures 5.2–5.5. Located on a base with good thermal conductivity (copper alloy) is a beryllium oxide layer, electrically isolated but likewise with good thermal conductivity. Applied on this is high-doped n^+ conductive silicon, with a thickness of about 200 μm, which forms the collector. Applied onto this is a thin (about 20 μm) weakly doped n^- epitaxial layer; it is this, in conjunction with the diffused p-conductive base layer, which determines the voltage characteristics of the transistor (level of the breakdown voltage BV_{CB}). The last doping (again, n-conductive) now forms the emitter. This npn structure is selected because of the higher mobility of the electrons for HF power transistors (Figure 5.2). To produce the power required the transistor consists of a large number of individual transistors connected in parallel (Figure 5.3). The structure by which they are connected together must be selected in such a way that the most uniform current distribution possible is produced across the entire chip.

For the development of the transistor the task is, among other things, to attain an emitter structure of such a nature that it will feature the greatest possible emitter edge length with a given emitter surface, since the injection of minority carriers into the base, at high frequencies, takes place essentially along the edge of the emitter (Figure 5.4).

After a number of experiments with overlay and emitter grid layouts, the step was taken in about 1970 by Valvo towards today's VHF/UHF power transistor technology with the invention of what is known as the 'beady structure' (Figures 5.4 and 5.5). The emitter resistance zone, which is diffused, provides for stable operation of the transistor. With the emitter current concentrated on one or only a few of the emitter strips, a voltage drop occurs over the emitter resistance, which counteracts the current concentration. Despite this series resistance in the emitter feed lines of the individual transistors, the load-dependent temperature distribution is derived as in Figure 5.6.

The connections of base and emitter to the outer connection strips must be provided by multiwired bonds with as little inductance as possible. The use of additional metal-oxide capacitors can achieve a preliminary matching with the help of the inevitable base-bond wires as inductance (Figure 5.5).

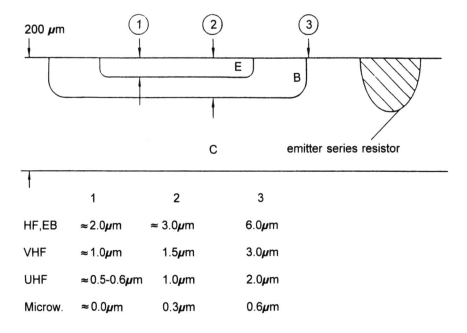

	1	2	3
HF,EB	$\approx 2.0\mu m$	$\approx 3.0\mu m$	$6.0\mu m$
VHF	$\approx 1.0\mu m$	$1.5\mu m$	$3.0\mu m$
UHF	$\approx 0.5-0.6\mu m$	$1.0\mu m$	$2.0\mu m$
Microw.	$\approx 0.0\mu m$	$0.3\mu m$	$0.6\mu m$

Figure 5.2   *Typical values of diffusion depths for different frequency ranges*

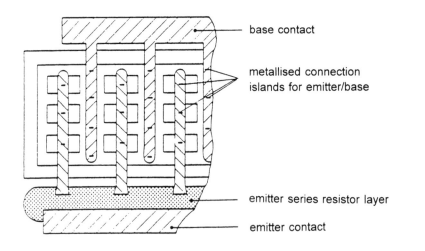

Figure 5.3   *Arrangement known as the 'beady structure': emitter sectors are connected to one another and to the contact via emitter resistors*

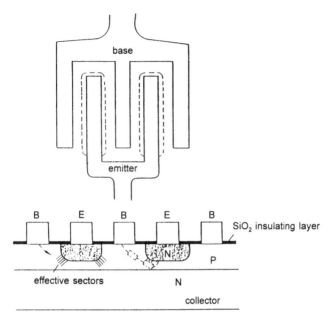

Figure 5.4   *Interdigital (comb) structure of HF transistor*

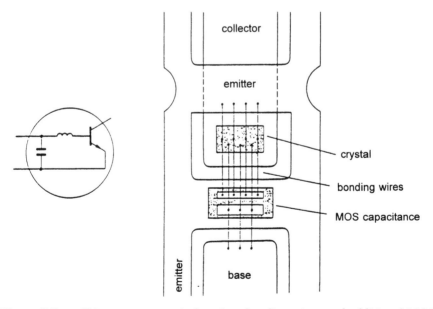

Figure 5.5   *Chip arrangement in housing; bonding wires and additional MOS capacitance provide preliminary adaptation for base–emitter link*

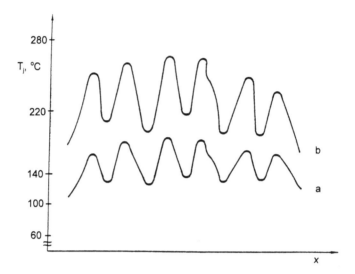

Figure 5.6   *Temperature distribution for different output loads*
(a) VSWR$_{load}$ = 1.1; *(b)* VSWR$_{load}$ = 3:1

5.2.1.1 Technical operating data

The most important details as far as the engineer is concerned for operating a VHF power transistor are now discussed, based on the data sheet 'HF/VHF power transistor BLW 78' from Philips, and Figures 5.7–5.16.

HF/VHF Power transistor

BLW 78 is an npn silicon planar epitaxial transistor intended for use in class A, AB or B operated mobile, industrial and military transmitters in the HF and VHF bands. It is resistance stabilised and is guaranteed to withstand severe load mismatch conditions. It has a 1/2" flange envelope with a ceramic cap. All leads are isolated from the flange.

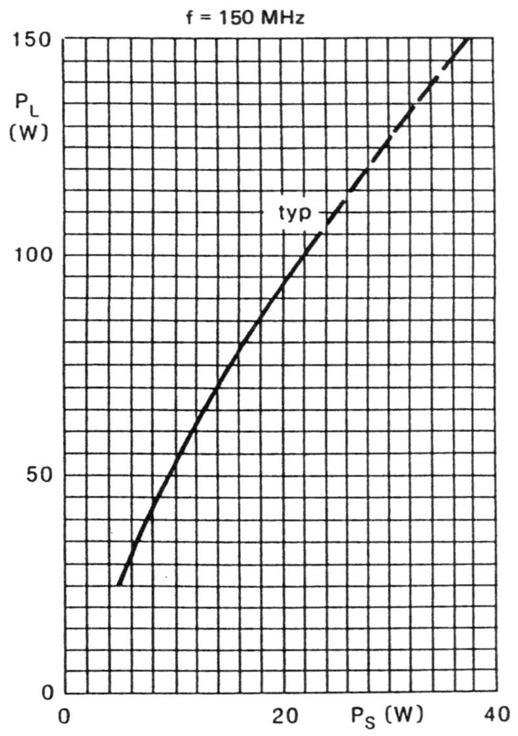

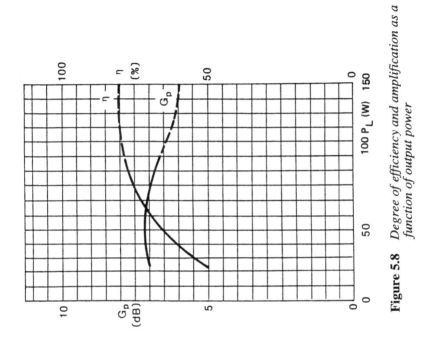

Figure 5.8   *Degree of efficiency and amplification as a function of output power*

Figure 5.7   *Control characteristics*

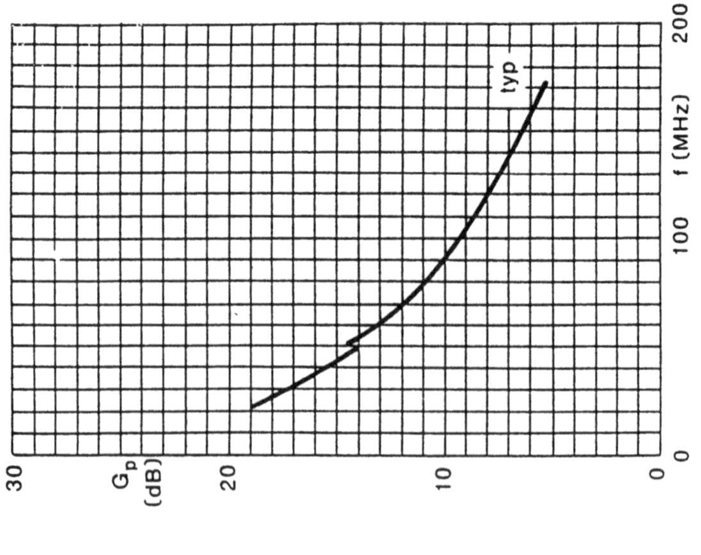

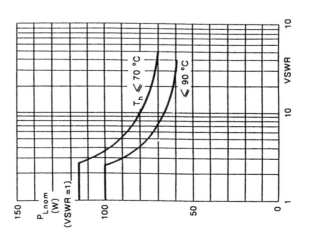
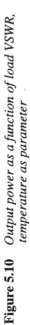

Figure 5.10   *Output power as a function of load VSWR, temperature as parameter*

Figure 5.9   *Amplification as a function of frequency*

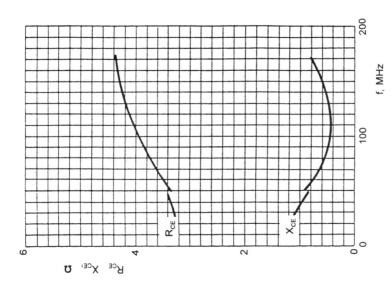

Figure 5.12 *Load impedance Z_CE (series connection)*

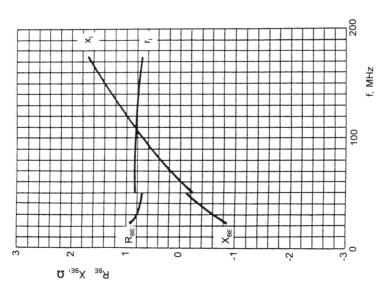

Figure 5.11 *Input impedance Z_BE (series connection)*

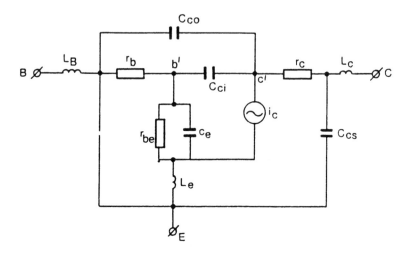

Figure 5.13   *Equivalent circuit diagram of bipolar HF power transistor*

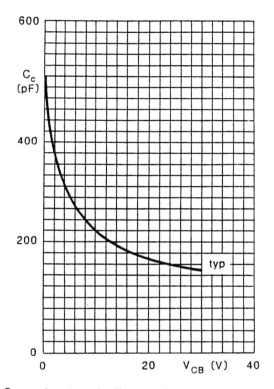

Figure 5.14 C_{ob} *as function of collector voltage*

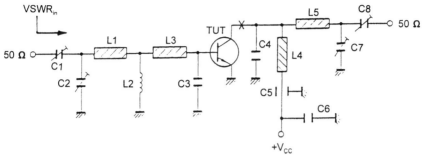

Figure 5.15   *Test circuit*

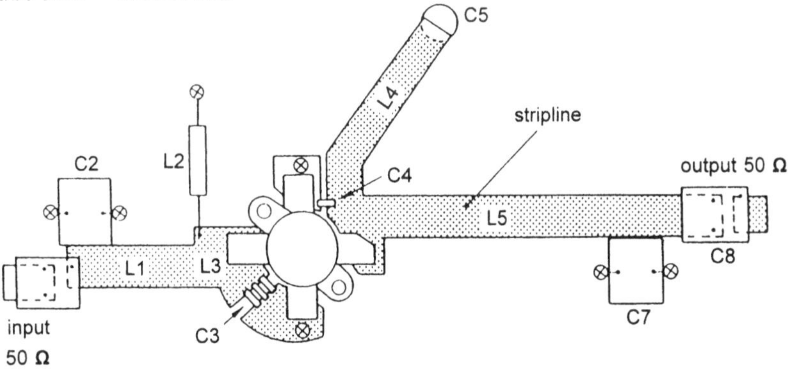

Figure 5.16   *Stripline arrangement of test circuit*

Quick reference data

RF performance up to $T_h = 25°C$

mode of operation	V_{CE} [V]	I_{qx} [A]	f [MHz]	P_L [W]	G_p [dB]	η [%]	$d3^*$ [dB]	Z_i [Ω]	Z_L [Ω]
CW (class B)	28	–	150	100	>6	>70	–	0.74 + J 1.35	4.3 + j0.60
SSB (class A)	26	3	28	35 (P.E.P.)	typ 19.5	–	typ – 40		
SSB (class A-B)	28	0.05– 0.10	28	100 (P.E.P.)	typ 19.0	typ 42	typ – 30		

* Intermodulation distortion figures are referred to the relevant power level of either of the
equal amplified tones. Relative to the relevant peak envelope powers these figures should
be increased by 6 dB.
I_{qx}: quiescent current

Characteristics

$T_j = 25°C$

Collector–emitter breakdown voltage		
$V_{BE} = 0$, $I_C = 50$ mA	$V_{(BR)CES}$	$>$ 70V
Collector–emitter breakdown voltage		
open base; $I_C = 100$ mA	$V_{(BR)CEA}$	$>$ 35 V
Emitter–base breakdown voltage		
open collector; $I_E = 5$ mA	$V_{(BR)EBO}$	$>$ 4V
Collector cut-off current		
$V_{BE} = 0$; $V_{CE} = 35$ V	I_{CES}	$<$ 5 mA
DC gain*		
$I_C = 5$ A; $V_{CE} = 5$ V	h_{FE}	20 to 85
Collector–emitter saturation voltage		
$I_C = 15$ A; $I_B = 3$ A	V_{CEsat}	typ. 2 V
Transition frequency at $f = 100$ MHz		
– $I_E = 5$ A; $V_{CB} = 28$ V	f_T	typ. 370 MHz
– $I_E = 15$ A; $V_{CB} = 28$ V	f_T	typ. 350 MHz
Collector capacitance at $f = 1$ MHz		
$I_E = I_e = 0$; $V_{CB} = 28$ V	C_c	typ. 155 pF
Feedback capacitance at $f = 1$ MHz		
$I_C = 100$ mA; $V_{CE} = 38$ V	C_{re}	typ. 102 pF
Collector–flange capacitance	C_{cf}	typ. 3pF

Thermal resistance:

Junction–mounting base	$R_{th\ j–mb}$ (dc) = 1.45 K/W (for DC load)
Junction–mounting base	$R_{th\ j–mb}$ (rf) = 1.06 K/W (for HF operation)
Mounting base–heatsink	$R_{th–mb–h} = 0.2$ K/W

It can be seen from the quick reference data that the transistor can output an HF power of 100 W at 28 or 150 MHz at a supply voltage of 28 V (Figure 5.7). The power amplification G_p at 150 MHz in this situation amounts to at least 6 dB; i.e. for a required output power of 100 W we need about

$$P_{st} = \frac{100\ W}{10^{\frac{6\ dB}{10}}} = 25\ W$$

With correct dimensioning of the output network, we obtain an efficiency of $\eta = 70\%$, where

$$\eta = \frac{\text{HF output power}}{\text{absorbed DC power}} = \frac{P_o}{P_i}$$

where $P_i = U_{CE}\ I_c$ (Figure 5.8).

At 28 MHz, the amplification according to Figure 5.9 rises sharply (approximately 19 dB) and wideband operation requires measures to linearise the frequency response. Special circuit-engineering tricks are then needed to prevent unwanted oscillation being generated at low frequencies.

Figure 5.10 shows an important diagram: the maximum output power at high temperatures and different load resistances $Z_{CE} = R_{CE} \pm jX_{CE}$, expressed by the $VSWR_{load}$ ($Z_0 = 50$ W) at the output port of the test circuit (Figure 5.15). The full output power of 100 W can, for example, still be produced at $T_h = 90°C$ and a mismatch $VSWR_{load} = 25{:}1$.

Figures 5.11 and 5.12 show typical frequency-dependent input and load impedances, as calculated from the equivalent diagram (Figure 5.13). More will be said about this in the section about matching.

An important factor for stability in a circuit is the voltage-dependent collector base capacitance C_{ob}. This is shown in Figure 5.14 as a function of the collector voltage.

The test circuit shown in Figure 5.15 is designed to acquire various measurement data, and, above all, to check it. The input network NW_1 is adjusted with the help of C_1 and C_2 in such a way that $VSWR_{IN} = 1{:}1$. By the appropriate adjustment of C_1 and C_2 at the input network, the circuit is balanced to maximum output power and good degree of efficiency. For the output arrangement, two so-called 'balance extremes' are possible. Adjustment to maximum amplification

$$G_p\ (dB) = 10\ \log \frac{P_o}{P_{st}}$$

or best efficiency $\eta = P_o/P_i$ (incorporating the harmonic energy and the effects of C_{ob}). In practice, the network is dimensioned between these two extremes. Figure 5.16 shows the practical arrangement of a test circuit of this type in stripline technology. The inductances $L_{1/2/4/5}$ are achieved by means of short line sections (striplines); the fixed capacitances ate usually compact component elements (high-value ceramics).

5.2.1.2 Operating point adjustment

Consider the test circuit according to Figure 5.15 and note that the base of the transistor for direct voltage over L_3 and L_2 is connected directly to earth. The collector current can therefore only flow if a satisfactorily large control power ($P_{st} > 0.3$ W) is provided at the input. As known from valve technology, in this case too the adjustment of the working point can be attained with class C operation, with which the best efficiency (theoretically 78%) can be achieved. The current flow angle θ in this case is < 90° with the result that linear amplification of modulated control signals cannot take place. The class C operation is therefore predominantly set for FM power stages.

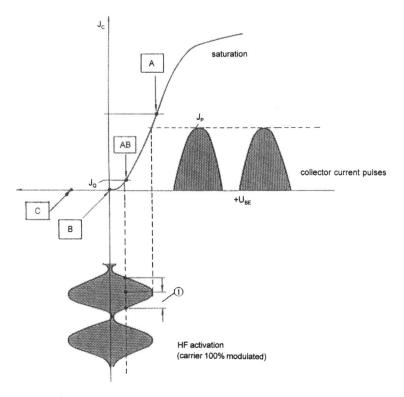

Figure 5.17 *Operating points*

If it is intended that a high-frequency power stage should work in linear form (AM, SSB, or television application), then either class A operation (maximum theoretical efficiency 50%) or what is known as class AB operation is required. In practice, of course, in class A operation the highest possible linearity may be achieved, but an efficiency of 50% is never attained (about 20%). Only the AB setting offers the compromise between good linearity and acceptable efficiency (40–60%). For a 100 W output stage, the quiescent current I_Q is to be set about 50 mA. In this situation, the power amplification is about 1 dB greater than in class C operation, up to some 70% of the maximum output power. The current source for the U_{BE} bias voltage of approximately 0.7 V must be very low ohmic for the working point not to be changed by HF rectification. In practice, the regulated generation of bias voltage has proved its worth. The temperature compensation required of U_{BE} can in this context be effected by an additional temperature measuring diode D_1 (Figure 5.18).

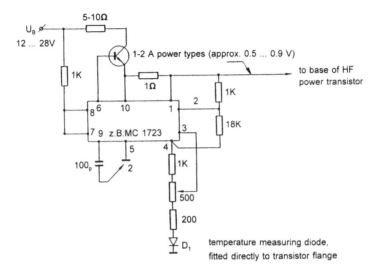

Figure 5.18   *Control circuit for production of low-ohmic B–E bias voltage with temperature compensation*

5.2.1.3 Thermal conduction

In view of the fact that no converters of supply power to high-frequency power (which includes the active transistor) operate without loss, one must take into account the fact that thermal energy will be created. This causes the transistor to heat, and, unless sufficient provision is made for the dispersal of the heat, the chip may be destroyed at an excess temperature $T_j > 200°C$, since the positive temperature coefficient leads to 'thermal runaway'. One can define the temperatures which occur with the help of an example (Figure 5.19).

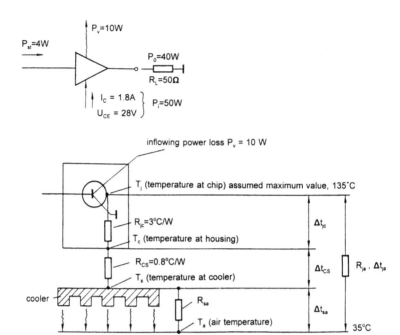

Figure 5.19 *Temperature distribution area surrounding chip*

An output stage produces an output power of 40 W at 144 MHz and at 50 Ω. The supply voltage is 28 V, and an I$_C$ flows of 1.8 A; i.e. the collector DC input power is 50 W, and the efficiency η = 80 % (with very low-loss networks, and special dimensioning of the output network, high efficiencies of this nature can be achieved; this is known as class E operation). The maximum permissible chip temperature is assumed to be 135°C, and the ambient temperature 35°C. In this case, the power loss P$_V$ = P$_i$ – P$_o$, by analogy, is regarded as heat, which flows through various thermal 'resistances' connected in series, and produces the corresponding 'voltage drops', Δ$_t$ (°C). The Δ$_t$ between chip and environment is therefore

$$\Delta t_{ja} = T_j - T_a = 135 - 35 = 100°C$$

and from this one can calculate a total thermal resistance of

$$R_{ja} = \frac{\Delta t_{ja}}{P_V} = \frac{100°C}{10W} = \frac{10°C}{W}$$

R$_{jc}$ is quoted by the manufacturer as 3°C/W, and the transition from the transistor housing to the cooling element R$_{jc}$ can be assumed to be 0.8°C/W. What size, and therefore what thermal resistance, must the cooling element have?

$$\Delta t_{jc} = R_{jc} \cdot P_v = 3 \, \frac{°C}{W} \cdot 10 \, W = 30°C; \ T_c = 135°C - 30°C = 105°C$$

$$\Delta t_{cs} = R_{cs} \cdot P_v = 0.8 \, \frac{°C}{W} \cdot 10 \, W = 8°C; \ T_s = 105°C - 8°C = 97°C$$

$$R_{sa} \frac{\Delta t_{ja}}{P_v} = \frac{97°C - 35°C}{10 \, W} = 6.2 \, \frac{°C}{W}$$

If an effective cooling element (larger surface area) is selected with a substantially smaller thermal resistance, e.g. 2°C/W, one obtains the following temperatures at the chip, with consistent power loss and ambient temperature:

$$R_{ja} = R_{ja} + R_{cs} + R_{sa}$$

$$= 3 \, \frac{°C}{W} + 0.8 \, \frac{°C}{W} + 2 \, \frac{°C}{W} = 5.8 \, \frac{°C}{W}$$

$$\Delta t = 5.8 \, \frac{°C}{W} \cdot 10 \, W = 58°C$$

$$T_j = 35°C + 58°C = 93°C$$

At 55°C, it is no longer hazardous to touch the chip. From the example, the thermal resistance of the chip environment must be made as small as possible. Conventional thermal resistances for chip housings are about 0.5°C/W for 100 W transistors. The heat transference from transistor housing to cooling element can be improved by broad-area smooth surfaces, using thinly applied thermally conductive paste. With power amplifiers > 100 W, most straightforward convection cooling elements are no longer adequate; in this case, forced cooling must be used by means of fans.

5.2.2 Field-effect transistors

For low power values field-effect transistors, FETs for short, predominantly feature the layout shown in Figure 5.20. The current flows, in what is known as the enhancement mode type, from the source to the drain when an appropriately high positive voltage is imposed at the gate relative to the source. The gate (polysilicon material) in this situation is isolated by an oxide layer, and therefore controls the charge (Figure 5.21). This current flow does, however, create a substantial voltage drop in the drain–source channel, since the channel resistance with this lateral technology is relatively high (1–5 Ω).

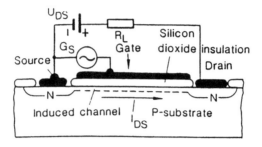

Figure 5.20 *Layout of MOSFET with horizontal channel arrangement (planar structure)*

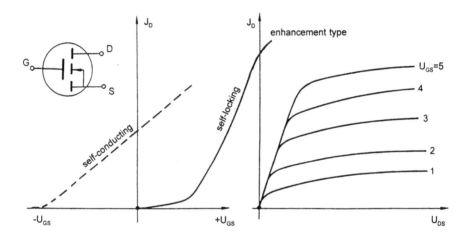

Figure 5.21 U_G/I_D *MOSFET characteristics*

The connections required on the upper side of the chip mean that large surface areas are required, which in turn increase the capacitance. In brief, a technology of this kind is not suitable for power FETs. The real breakthrough for FETs into the high-frequency power sector first came about with the possibility of the vertical current flow (Figures 5.22 and 5.23). Manufacturers rapidly developed a series of processes (VMOS, TMOS, DMOS) in which the current flows from top (source) to bottom (drain). This means that the heat loss can be dissipated, the channel sections can be shortened again and made lower-ohmic by means of parallel connections (R_{DS-ON} approx. 0.5 Ω) and the capacitances smaller.

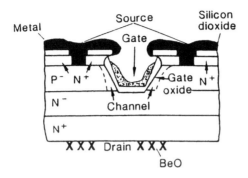

Figure 5.22   *Vertical current flow in VMOS-FET*

The vertical arrangement allows for a cell-shaped structure of the power semiconductor to be established, and therefore permits a high concentration of identical MOSFET elements on the crystal. Figure 5.23 illustrates a DMOS transistor (Philips). Housing and bond technique are of the same design as in bipolar technology. On the heat sink (BeO + copper contacts) is a low-ohmic n^+substrate, and on this the n-region (an epitaxial layer, which provides for high breakdown voltage $V_{(BR)DSS}$ and small capacitance). The current starts from the n^+source region, which is diffused, then continues through the p-region (channel) to the drain. This current can be controlled by the voltage imposed at the gate. The channel width in this context can be seen vertically on the drawing. This is required for the maximum current and output power (R_{DS-ON} steepness) and linearity at high currents.

The feedback capacitance is very small owing to the oxide layer applied beneath the gate, which is important for stability in the case of high steepness values. The layout sketch shows the three most important undesired capacitances C_{DS}, C_{GD}, C_{GS} and the current path I_D. C_{GD} in this situation is composed of a fixed and a voltage-dependent capacitance (displacement of the depletion area), C_{DS} is a pn capacitance, and therefore, as with the bipolar transistor, voltage-dependent. Only C_{GS} is to be regarded as constant.

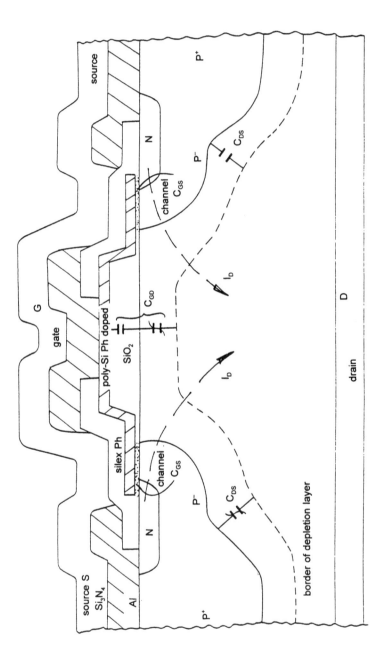

Figure 5.23 *Schematic diagram of DMOS-FET (double diffused)*

The measurement capacitances are given by the manufacturer (Figure 5.24), which are related to the internal capacitances as $C_{iss} = C_{GD} + C_{GS}$, $C_{oss} = C_{GD} + C_{DS}$ and $C_{rss} = C_{GD}$.

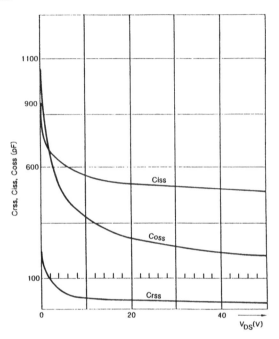

Figure 5.24   *Curve of capacitances C_{rss}, C_{iss}, and C_{oss} against drain-source voltage imposed (measured value per transistor section, $T = 25°C$)*

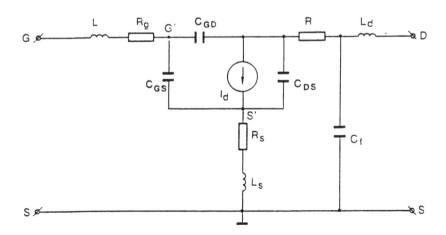

Figure 5.25   *Equivalent circuit of HF power MOSFET*

From the layout sketch (Figure 5.23) and the equivalent circuit (Figure 5.25), it can be seen that the input impedance Z_{GS} at low frequencies (up to about 100 MHz) consists of a series circuit, a capacitance, and a very small loss resistance (R_g, R_s). As a first approximation, the MOSFET is, by contrast with the bipolar transistor, a voltage-controlled amplifier component.

The higher the operating frequency becomes, the more predominant the inductive component, and, in part, also the ohmic component becomes (Figure 5.26). For very large frequency ranges, then, we have a somewhat more complicated matching arrangement than with the bipolar transistor. The higher amplification with the MOSFET can no longer be fully utilised because of the additional frequency response linearising resistors which are required.

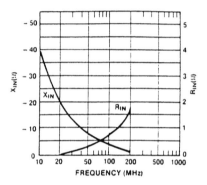

Figure 5.26a *Curve of input impedance as a function of frequency of 100 W MOSFET for 100–200 MHz*

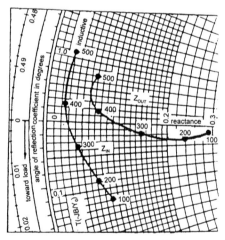

Figure 5.26b *Curve of input and output impedance as a function of frequency of 100 W MOSFET for 100–500 MHz*

The curve of the load resistance R_{DS} required is similar to that for the bipolar transistor. The output capacitance C_{oss}, which in most cases is somewhat smaller, has a favourable effect for wideband applications.

5.2.2.1 VHF push-pull power MOS transistor

BLF 278 is a push-pull silicon n-channel enhancement mode vertical D-MOS transistor intended for use in VHF broadcast transmitters. The transmitter has a four-lead balanced flange envelope with two ceramic caps. The flange is the source.

Quick reference data
RF performance at $T_h = 25°C$ in a push-pull common-source class-B test circuit

Mode of operation	f [MHz]	V_{DS} [V]	P_L [W]	G_D [dB]	η [%]	J_{DQ} [A]
CW (AB)	108	50	300	> 18	> 60	2 x 0.1 A
CW	175	50	300	typ. 16	typ. 55	

$P_{tot} = 500$ W, $V_{DSmax} = 110$ V

Characteristics $T_h = 25°C$

Drain–source breakdown voltage
    $V_{GS} = 0$, $I_D = 50$ mA $V_{(BR)DSS}$ > 110 V

Drain–source leakage current
    $V_{DS} = 50$ V, $V_{GS} = 0$ V I_{DSS} < 2.5 mA

Gate–source leakage current
    $\pm V_{GS} = 20$ V, $V_{DS} = 0$ I_{GSS} < 1 μA

Gate threshold voltage
    $I_D = 50$ mA, $V_{DS} = 10$ V $V_{GS(th)}$ 2.0–4.5 V

Gate threshold voltage diff.
of both transistor sections
    $I_D = 50$ mA, $V_{DS} = 10$ V $V_{GS(th)}$ < 100 mV

Forward transductance $G_f =$ > 4.5 S
    $V_{DS} = 10$ V, $I_D = 5$ A   typ. 6.2 S

Drain–source ON-resistance
    $I_D = 5$ A, $V_{GS} = 10$ V $R_{DS(on)}$ typ. 0.2 Ω

On-state drain current
    $V_{DS} = 10$ V, $V_{GS} = 10$ V I_{DS} typ. 27 A

Input capacitance
    $V_{DS} = 50$ V, $V_{GS} = 0$, $f = 1$ MHz C_{iss} typ. 470 pF

Output capacitance
    $V_{DS} = 50$ V, $V_{GS} = 0$, $f = 1$ MHz C_{oss} typ. 190 pF

Feedback capacitance
    $V_{DS} = 50$ V, $V_{GS} = 0$, $f = 1$ MHz C_{rss} typ. 19 pF

Operating junction temperature T_j max. 200°C

Thermal resistance (total device)

Junction to mounting base $R_{th\ mb-h} = 0.35$ K/W

Mounting base to heatsink $R_{th\ mb-h} = 0.15$ K/W

Load mismatch
The BLF 278 is capable of withstanding full load mismatch (VSWR = 7 through all phases) under the conditions: $V_{DS} = 50$ V, $f = 108$ MHz, $T_h = 25$°C, $R_{th\ mb-h} = 0.15$ K/W at rated output power.

Technical operating data
With a supply voltage of 50 V, a power value of 300 W can still be produced at 175 MHz. Depending on the complexity at the gate–gate match, an amplification of up to 20 dB is possible in this situation (narrowband operation). The relatively small feedback capacitance of 19 pF provides for high stability. It is nevertheless appropriate to provide attenuation resistors at the gates. In a similar way to the situation with bipolar transistors, the MOSFET is also still usable for high-load VSWR. The data given under load mismatch indicates that the transistor can be switched from normal operation ($R_L = 50$ Ω, 300 W) to a load with VSWR = 7 without a defect occurring in the chip. The housing must, however, remain at 25°C. Figure 5.27 shows the relation between output power and amplification, and the temperature of the housing. The steepness drops somewhat as the temperature rises, which is indicated by the dashed curves for P_o and G_P; the difference is about 0.5 dB.

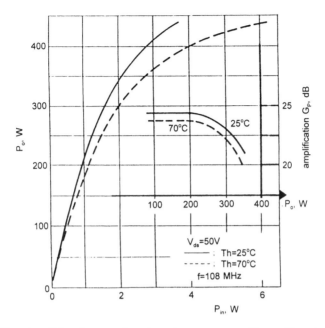

Figure 5.27 *Amplification as a function of output power P_O and housing tempera-ture: Output power as a function of input power at 25 and 70°C housing temperature, load resistance of emitter 3.7 + j 3.8 Ω.*

5.2.2.2 Operating point adjustment

In view of the fact that the semiconductor element is of the self-locking type, a positive bias (> $U_{GS(th)}$, approximately 2 to 5 V) must be fed to the gates to attain, for example, a quiescent current J_{DQ} of 0.15 A. By varying this bias, the HF output power can be altered within broad limits, at constant input power. Without any positive U_{GS}, the steepness is very slight; for this 'super class C operation', a very high input amplitude would be required. The usual kind of operation used with a MOSFET, then, is B-AB-B class operation. By contrast with the bipolar transistor, the MOSFET has a negative temperature coefficient; in other words, the transistor does not undergo thermal drift. Regulation of the gate bias by a temperature sensor diode on the HF transistor housing is not required.

5.2.3 Comparison of MOSFET and bipolar transistors

Advantages of MOSFET

- Thermal stability thanks to a negative temperature coefficient; this prevents local overloads (hot spotting, secondary breakdown)

- High input impedance over a large frequency range (kΩ) almost purely capacitative; voltage control

- Slight change in input impedance with control; no feedback

- Higher amplification by 2–3 dB than bipolar transistor

- Straightforward direct voltage generation at the gate for I_{DQ} and gain control by means of V_{GS}

- Low noise due to no effective pn transitions (improvement by about 3 dB)

- Higher working voltage (50–100 V), greater linearity leads to lower intermodulation products of a higher order (resulting in lower interference from adjacent channels)

Disadvantages of MOSFET

— Careful handling at the gates; static charges may damage the oxide layer beneath the gate surface at an early stage; incorporate the transistor as the last stage in the complete circuit

— Higher amplification feedback (steepness) leads to a power drop as temperatures rise

— Complicated input networks at very high bandwidths

— Greater sensitivity for reflected harmonics in the output network (stability)

— Higher price than comparable bipolar transistors

5.3 Circuit engineering

5.3.1 The single-ended amplifier

Figure 5.28 shows the principal circuit diagram of a single-stage bipolar HF power amplifer with the measured values corresponding to actual practice. Depending on the bandwidth required, the matching networks consist only of simple L/C circuits or of more extensive multielement L/C circuits in connection with wideband power transformers. Their task in general is to transform the complex input impedance Z_{BE} to $R_e = 50\ \Omega$, or the load resistance $R_L = 50\ \Omega$ to the required load resistance Z_{CE} in the frequency range required. As the example shows, these networks are unfortunately not loss-free, and this must be taken into account in planning the overall amplification and output power.

In view of the fact that the transistor is a nonlinear component, harmonics are created in addition to the amplified fundamental wave, especially in C class operation. These, of course, should not be emitted; in other words, then, a harmonic filter (lowpass) must be looped in on the route to the antenna. It is also possible to eliminate the energy of the harmonics in an absorber filter (diplexer).

At the terminal + U_{BE} the appropriate bias can be imposed (0 V for C class operation, 0.3–0.8 V for A or AB class operation). The voltage input networks for the base (L_1, Fe, L_2, R_1) and collector (C_1, R_2, C_2, R_3, L_3, L_4) call for particular attention with regard to stability (uncontrolled oscillation, so-called parametric effects).

A three-stage wideband amplifier, likewise bipolar (100–156 MHz) for the air navigation band is shown in Figure 5.29, from Rohde & Schwarz. In this case, almost all the networks are formed from pure L/C circuits. The load impedance $R_L = 50\ \Omega$ is brought to the load resistance required Z_{CE3} by means of a transforming lowpass filter. The matching of Z_{BE2} to Z_{CE1} is effected with the aid of the network N_{W1} and the wideband transformer Tr_1 ($U_R = 4{:}1$). The input impedance Z_{BE1} of T_1 is loaded with an attenuator (α_0 approximately 1 dB) in addition to the L/C network. Thanks to this additional attenuation at the input, the overall amplification is indeed reduced by 1 dB, but the feedback stability to the preceding amplifier stages, e.g. in frequency processing, is substantially increased.

To correct the frequency response use is made, among other things, of the frequency-dependent voltage negative feedback amplifier incorporated in the first stage (R_1, C_0, R_2). In principle, the intermediate networks of Z_{CE1} are dimensioned to Z_{BE2} and Z_{BE3} in such a way that the best matching is located at the upper end of the transmission frequency band. The rising amplification of the transistors in the lower transmission range is then offset by the matching becoming increasingly worse towards the low frequencies.

5.3.1.1 Matching networks

The layout of a transistor power amplifier is, in practice, essentially a problem of matching. This involves the transformation of a 50 Ω load or generator resistance to the relatively low output and input impedances, or the matching of the input impedances of a transistor stage to the load resistance required, which is likewise complex, for the primary stage (Figure 5.29).

From the equivalent circuit diagram (Figure 5.13), and in particular from the data provided by the manufacturer, a low-ohmic real part with series inductance can be provided for the input, and a real resistance with a parallel or series capacitance can be produced as an equivalent circuit for the output (Figures 5.30a and 5.30b).

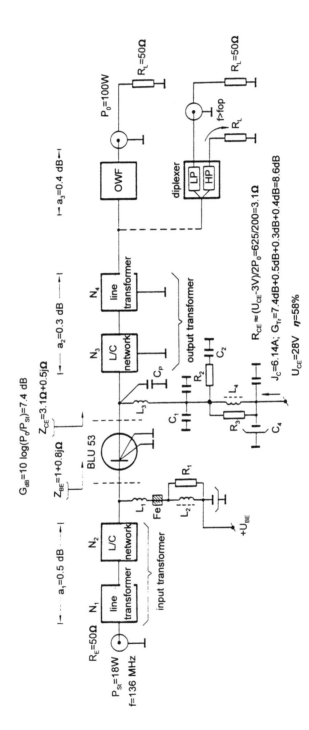

Figure 5.28   *Principal circuit diagram of high-frequency power stage for VHF with* $P_o = 100$ *W*

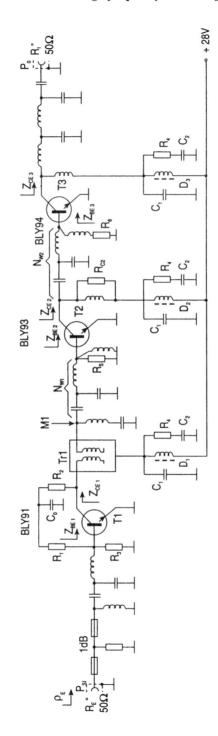

Figure 5.29   *Three-stage VHF amplifier for air navigation band 100–150 MHz, with output power of $P_o = 50$ W; total amplification approx. 26 dB*

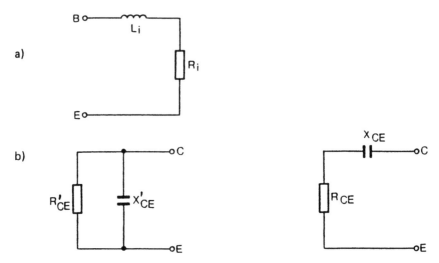

Figure 5.30 *Equivalent circuits for base–emitter section or collector–emitter section*

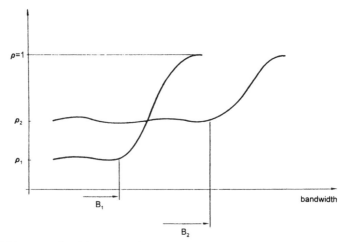

Figure 5.31 *Bandwidth/input reflection*

The quality to be selected for the networks depends on the transmission bandwidth required. Narrowband circuits have Q values of between 7 and 15, while wideband stages have Qs between 1 and 3. The question of whether a transistor is suitable for wideband operation depends primarily on the ratio $\text{Im}\,(Z_{BE})/\text{Re}\,(Z_{BE})$. The higher the series inductance of the complex base emitter impedance, the lower the bandwidth attainable, even with an infinitely large number of components in the matching network. A small input VSWR will only be attained in a small frequency range, and vice versa. The connections between theoretical minimum attainable input VSWR at a given bandwidth and $Z_{BE} = R_e + wL_e$ provides what is known as the Fano integral:

$$\int_0^\infty \ln/\frac{1}{\rho E} \, d\omega = \pi \cdot \frac{R_e}{L_e}$$

$$|\rho E|_{min} = \exp\left[\frac{-\pi \cdot R_e}{(\omega_0 - \omega_u)L_e}\right]$$

An example should serve to explain this. If the values for L_e and R_e are inserted, then $\rho E = 0.206$; this corresponds to an input VSWR = 1.5 : 1. If the control voltage offered amounts, for example, to 1.0 W, 0.2 W will be reflected. The transistor then receives a real control power of 0.8 W.

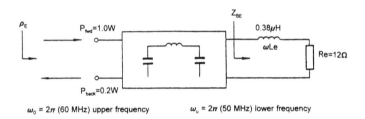

$\omega_0 = 2\pi$ (60 MHz) upper frequency $\omega_u = 2\pi$ (50 MHz) lower frequency

Figure 5.32   *To determine minimum input reflection according to Fano*

A straightforward graphical determination of the switch elements of a matching network, with coils and capacitors, can be carried out with the aid of the admittance diagram or Smith diagram, with the Q circles drawn in.

This is demonstrated in an example (Figure 5.33), in which the values for L_1, C_1, L_2 and C_2 are determined for an output network. The circuit is designed in two stages, to achieve the desired Q of 2. For the real intermediate value, the geometrical mean is chosen:

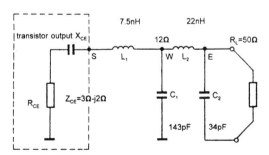

Figure 5.33   *Output circuit with L/C transformation*

$$W = \sqrt{Re \cdot Z_{CE} \cdot R_L} = \sqrt{3 \cdot 50} \approx 12 \, \Omega$$

The transistor output resistance $Z_{CE} = 3 - j \, 2 \, \Omega$ is drawn in at the point S as a normalised value ($Z_{CE}/50 \, \Omega$). The first inductive series impedance ωL_1 extends as far as the point of intersection M, and after W is transformed in real terms to $12 \, \Omega$ by means of the capacitative admittance value ωC_1. The section W–T represents ωL_2; rotation is effected to $50 \, \Omega$ (E) by means of ωC_2. The points of intersection M and T remain, as required, at $Q \approx 2$. The values from the graphic representation are the L/C values given in Figure 5.33. The matching circuit shown here approximates very closely to what are referred to as the 'transforming low-passes'.

To obtain the dimensions of very wideband matching networks of this type, the ratio of real output and input impedance is first determined:

$$r = \frac{R_L}{Re \, Z_{CE}}$$

and a standardised bandwidth value $W = (f_{max} - f_{min}) / (f_{max} + f_{min})$. On the basis of a table, and taking into consideration the prescribed maximum insertion loss, the number of components required n can be determined. In a further table, the values of the individual component elements can then be calculated from n and r.

If the appropriate manufacturer's data is not available, the real part (parallel equivalent circuit) of R'_{CE} can be roughly determined from the formula

$$P_o = \frac{U_{CE}^2}{2} \cdot R'_{CE}$$

By taking into account the knee-point voltage and the residual voltage U_K, we obtain for R'_{CE} (real part of the load impedance):

$$R'_{CE} = \frac{(U_{CE} - U_K)^2}{2P_o}$$

where U_{CE} is collector emitter supply voltage and U_K is knee-point voltage, approx. 2–5 V. In the first approximation, it is possible to offset the transistor output capacitance C_{CE} (composed of the collector-voltage dependent feedback capacitance C_{ob} and the stray capacitances) by the appropriate measurement of the collector voltage feed reactor (L_3 in Figure 5.28).

With wideband networks, however, it is only possible to achieve compensation which is satisfactory in the entire transmission frequency range by means of the appropriate redimensioning of the real transformed networks. For optimum matching (high efficiency), or if data for R_{CE} is not available, the $Z_{CE} = F(f)$ must be measured for all frequencies of interest with the help of a test circuit (Figure 5.15) and the aid

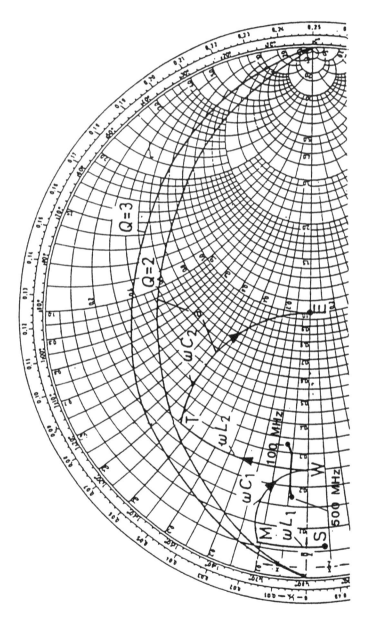

Figure 5.34   *Solution of matching task with the Smith chart, with circles for* Q = 2 *and 3.*

of a network analyser. Once the input network has been balanced to VSWR$_{in}$ = 1:1 and the output network to the desired output power P$_O$, then, for example at best efficiency, the connection between collector and output network will be separated at the defined point x. Z$_{CE}$ is now measured with a network analyser and entered on a Smith diagram. This somewhat exhaustive procedure must be carried out over the entire frequency range of interest.

With the help of modern computer programs (Super Compact, Touchstone), a network can then be optimised which will very closely approximate the required transformation. Entered in Figure 5.34 is the normalised curve of Z$_{CE}$ of a power transistor for the frequency range from 100–400 MHz. The maximum output power, with high efficiency, can of course only be obtained with the specified load of 50 Ω. Figure 5.35 shows how the output power drops (with constant input) if the load impedance L$_{CE}$ changes in the real and imaginary part (R$_L$ → R$_L$ ≠ 50 Ω).

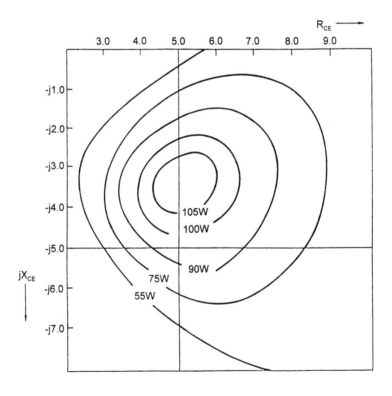

Figure 5.35   *Dependency of output power on complex load being supplied*

On the basis of Figure 5.36a, reference is made to a wideband transformation circuit with L and C known as the Meinke transformation. The transformation process is shown in Figure 5.36b. In the application considered here, R_1 represents the real part of the base–emitter resistance of Z_{BE}, the inductive part of which can now be incorporated in L_1. On the collector side, L_2 can be used as a feed voltage supply reactor.

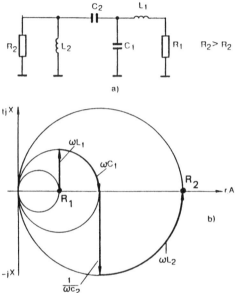

Figure 5.36 *Wideband transformation circuit according to Meinke*

Line transformers
Line transformers occupy a substantial place in modern transistor amplifier technology for large bandwidths (e.g. 1–200 MHz). By creating sensible layouts of one or more Guanella arrangements as shown in Figure 5.37, wideband transformers can be created with the transformation ratios 1:1, 4:1, 9:1, 16:1 and others. A two-wire lead Z_0 (strip or coax cable) with a length L < λ/8 at the highest frequency is wound

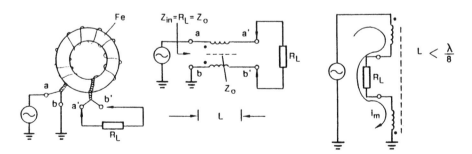

Figure 5.37 *Line transformer 1:1 after Guanella (suitable as a 180° phase rotator)*

onto an Fe core (even without Fe for smaller bandwidths and frequencies 100 MHz). The energy flow from generator to load is not affected by this. Because of the inductances formed between a and a' and b and b', however, asymmetrical flows are severely suppressed by single coil windings. On the primary a–b and the secondary a'–b' earthing can be applied at will; in other words, the input side is isolated from the output side. This makes it possible to connect input and output accordingly to obtain the transformation ratios described.

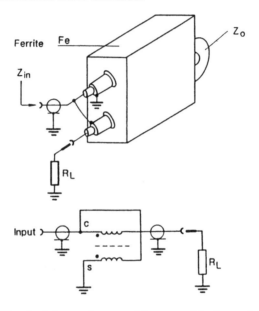

Figure 5.38   *Simple 4:1 transformer with coax cable through Fe block*

Figures 5.38 to 5.41 show some of the possibilities available. Depending on the frequency range and the power which is flowing (losses in the cable and Fe), wideband characteristics can be obtained up to 1000:1 with 4:1 transformers. Figure 5.39 shows an easily understandable arrangement of a 4:1 transformer. The voltage at the resistor R_1 (end of cable 1) is conducted via cable 2 to the primary side, and connected there in series to the input of cable 1. $U_1 = 2 \times U_2$, $U_u = 2$, and therefore a resistance ratio $U_R: = (U_u)2 = 4$.

In the main diagram (Figure 5.38), two line transformers are used. N_1 converts the 50 Ω generator resistance to 50 Ω/4 = 12.5 Ω. The L/C circuit that follows can only be set up more simply with a given wideband requirement, since it only has to transform 12.5 Ω to $1 + j\,0.8$ Ω. The same applies to the output: the load resistance $R_L = 50$ Ω becomes 12.5 Ω at the input of N_4. The LC network, which has once again become straightforward, is transformed to $3.1 + j\,0.5$ Ω. If there are several final stages, N_3 can almost be done away with altogether if the transformation in the

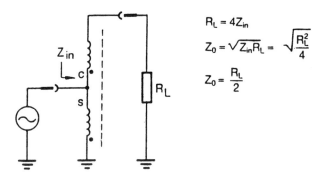

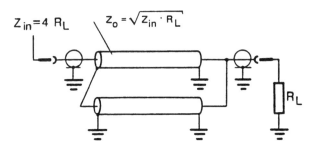

Figure 5.39   *4:1 transformer with phase compensation, asymmetrical to asymmetrical*

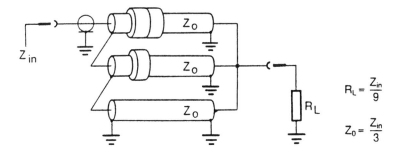

Figure 5.40   *8:1 transformer, asymmetrical to asymmetrical*

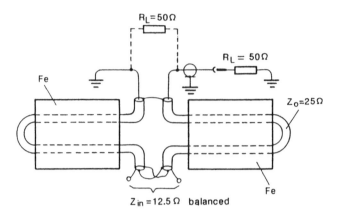

Figure 5.41 *4:1 transformer, symmetrical (50 to 12.5 Ω) for push-pull operation*

real part already matches, and only partial compensation of the imaginary part of Z_{CE} takes place (examples with the push-pull circuits).

5.3.2 Push-pull amplifiers

As with LF technology, the use of two equal transformers in a push-pull circuit is one solution for increasing the power of an HF amplifier. As a sensible solution, individual transistors with an output power of about 150 W can be connected together for frequencies of up to 200 MHz. A comparison of push-pull and parallel circuits of two transistors provides striking evidence of the advantages of the push-pull circuit (Figure 5.42).

Input and output impedances are about four times higher than with parallel wiring, and the transformation ratio at 50 Ω is substantially smaller. As a result, a substantially higher bandwidth is attained.

The shape of the housing determines the maximum for the transistors, and therefore the size of the input lead and the common emitter inductance. These inductance values may cause problems (with stability) at higher frequencies (> 300 MHz), and reduce the advantages listed of higher input and output imped-ances. For this reason, twin (gemini) transistors have been developed in the past few years for high bipolar and MOSFET output power values.

In this context, two individual semiconductors, with identical data as far as possible are housed in a common housing (see Figure 5.1). Base and collector contacts are designed to be of broad surface areas, the common emitter at the same time forming the housing base. Because of the base–base symmetrical excitation which is then possible, this is only of importance for the DC feed. The amplification of a gemini arrangement is about 1 dB higher than corresponding individual transistors in a self-built push-pull circuit.

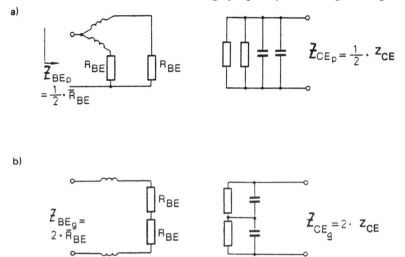

Figure 5.42 *Input and output impedances with parallel and push-pull circuits with two transistors*

The signals at the bases 1 and 2 are of equal amplitude, but phase-displaced to one another by 180°. The midpoint of the circuit (base and collector side) are at zero potential; in other words, a virtual earth. No groundwave current flows at the collector infeed. Decoupling is only necessary for the harmonics (see Figure 5.52). The suppression of even-numbered harmonics in a push-pull circuit depends on the symmetry balance of the circuit as a whole; this is highest with gemini transistors. The complexity (number of elements) of the harmonic filter is accordingly reduced.

Figure 5.43 shows a circuit with individual transistors for a 250–300 W wideband amplifier for single-sideband operation, frequency range 1.5–30 MHz.

To produce the output power desired, two high-voltage MOSFET DVD 120T (PHI) power transistors are used in a wideband circuit. A load resistance R_{DD} of 54 Ω can be determined approximately from the power and drain voltage values.

$$R_L = \frac{2(U_{DS} - U_{DK})^2}{P_o}$$

$$= \frac{2(100 - 10)^2}{300} = 54 \ \Omega$$

where U_{DS} = supply voltage (100 V), U_{DK} = knee-point voltage (10 V) and P_o = transistor power (300 W). For the two transistors operating in push-pull mode, a real load resistance of 56 Ω is required. The generally asymmetrical load of 50 Ω (coax output) can be provided symmetrically with the two drain connections with the aid of a straightforward 1:1 Guanella transformer. The voltage supply is effected

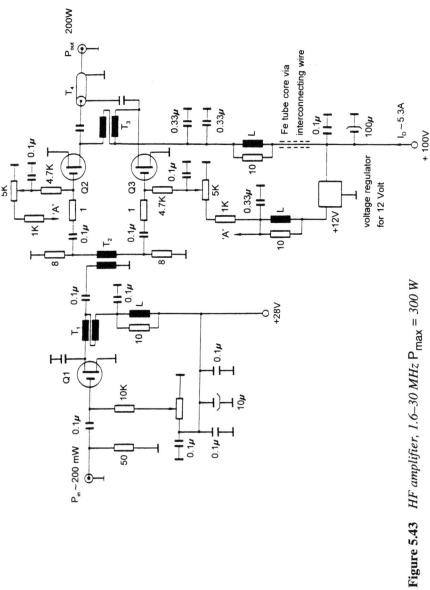

Figure 5.43   *HF amplifier, 1.6–30 MHz P*max *= 300 W*

via a 'coupled inductance', consisting of a 1:4 line transformer. This presents a short-circuit for all even-numbered harmonics.

On the input side, the conversion of the symmetrical input signal offered by the driver, of about 4 W takes place to the symmetrical gate–gate voltage required by means of a conventionally arranged 1:4 transformer (Figure 5.44)

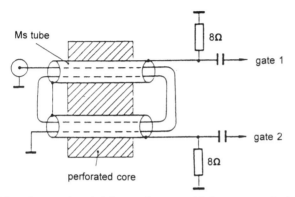

Figure 5.44   *Conventional 4:1 transformer with low scatter inductance*

The series resistors located in the gate input leads of 1 Ω do somewhat reduce the amplification, but they also provide for stable operating conditions. A straightforward voltage divider serves to adjust the corresponding DC gate voltage for the quiescent current required of about 0.2 A per transistor. By means of the isolated adjustment of the gate voltage it is possible, by means of careful balancing, to achieve a minimum of intermodulation products at individual operating frequencies. The input transformer T_2 is terminated on the gate side by $2 \times 8\ \Omega = 16\ \Omega$. With the gate circuits connected to this, a return loss of about 18 dB for good wideband matching is obtained. Without the additional real connection ($2 \times 8\ \Omega$), the amplification would be somewhat higher (about 1 dB), but the wideband properties would be reduced.

The control power required, of about 4 W, is supplied by a single-pulse MOSFET transistor stage with the transistor DV 2810S or similar. For this to be operated individually, it is dimensioned for an input and output impedance of 50 Ω. On the output side, a conventional 4:1 line transformer is used; depending on the transformation ratio, an R_{DS} of 200 Ω is provided for the drain–source section.

For the sake of simplicity, a wideband 50 Ω input is created with the aid of a 50 Ω resistor at the input. This substantially increases stability; the gain of the stage is still 13 dB as a minimum. Because the stage is operating for the lowest IMP in class A operation, a quiescent current of $I_{DQ} = 1$ A is to be set.

This two-stage HF amplifier produces a reliable output power of 200–230 W in the entire band which is of interest, with the intermodulation products being located at a spacing from a carrier (PEP = 220 W) of $2f_1 - f_2$ or $2f_2$ at –32 to –34 dB. The intermodulation products of higher order have a substantially greater interval, at 38–45 dB, than with bipolar transistors (Figure 5.45).

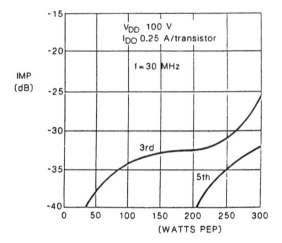

Figure 5.45 *Intermodulation products (IMP), as a function of output power/ distance spacing to carrier with two-tone measurement*

5.3.2.1 Matching networks

In every push-pull stage the task arises of converting the asymmetrical (unbalanced) control voltage which is introduced into an earth-symmetrical 180° offset voltage. In addition to this, of course, the input impedance R_{BB} or R_{GS} of the pair of transistors is to be transformed to the generator resistance (in most cases, 50 Ω).

For the first task, the Guanella transformer described earlier is particularly well suited. This has a transformation ratio of 1:1 ($R_G = R_L$), and divides U_1 into two equal values, rotated by 180° against the virtual earth. The transformation task of transforming the symmetrical input impedance, e.g. $Z_{BB} = 3 + 1.5\ \Omega$ to 50 Ω, can of course be resolved once again with a wideband L/C circuit. For very large bandwidths, however, it is a more sensible idea to divide the resistance transformation into two parts (Figure 5.38, L/C and symmetrical wideband transformer).

Figure 5.46 shows a 300 W power amplifier with the gemini transistor (MOSFET) BLF 278 from Philips. The load impedance $R_L = 50\ \Omega$ is converted symmetrically by means of T_4 to 50 Ω. T_3 transforms with a voltage ratio of 2:1, and from this there follows a resistance stepdown to 50:4 = 12.5 Ω. This is a usable real load resistance for the D_1–D_2 section. L_1 with C_1 and C_3 serves to provide compensation of the imaginary part at the output of the semiconductor. The input impedance Z_{GG}, which up to about 200 MHz has a purely capacitative effect, is produced by means of a similar arrangement T_1, T_2. By the appropriate selection of $R_{1/2/3}$, one can also speak of power matching, as with bipolar transistors. In this case, we are opting for a compromise between the maximum gate–gate voltage (amplification) and an adequate VSWR at the input (pay attention to ensuring stability). In view of the fact that the amplifier stage is supposed to operate in a linear fashion, a quiescent current must be set of about 0.6 A per transistor, with the aid of P_1 and P_2.

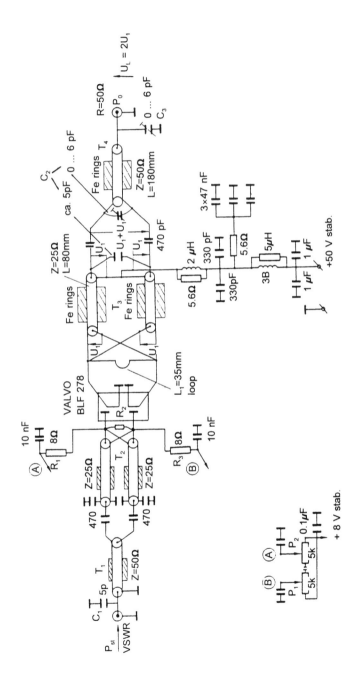

Figure 5.46   *300 W linear final stage for 70–160 MHz*

5.3.3 Stability

From the data for Z_{BB}, Z_{CE}, and in particular C_{ob}, all of which are input-dependent, one can see that the transistor is a nonlinear component. Its behaviour can be represented by two individual amplifiers wired in parallel, linear and nonlinear.

If we arrange, by means of an HF stage, with appropriate mechanical and electrical layout, that no external couplings occur, then one can determine the stability and conditions for the linear part from the S-parameters. It is self-evident that, in the case of multistage amplifiers, the decoupling of the individual stages takes place on the current supply side by means of the appropriate networks (D_1, D_2, D_3 with C_1, and R_4, see also Figure 5.29).

The input reflection coefficient $\rho_E = S'_{11}$ of a quadripole with termination at the output is obtained by

$$\rho_L = \frac{Z_L - Z_0}{Z_L + Z_0}$$

(where Z_L = complex load impedance; Z_0 = real reference impedance) and reduced to

$$S'_{11} = S_{11} + \frac{S_{21} \cdot S_{12} \cdot \rho_L}{1 - S_{22} \cdot \rho_L}$$

If coefficient S_{11} is great enough, and ρ_L is correct to phase, ρ_E can become > 1. HF energy then flows out of the input of the transistor; in other words, the arrangement oscillates. From the S-parameters measured, one can derive what are known as the stability conditions of K:

$$K = \frac{1 + |S_{11} \cdot S_{22} - S_{12} \cdot S_{21}|^2 - |S_{11}|^2 - |S_{22}|^2}{2|S_{12} \cdot S_{21}|}$$

With $K > 1$ the transistor is stable with all possible passive input and output terminations, i.e. any desired generator and load impedances may be connected. With the help of a graphical process (Figure 5.47) and $K < 1$, those generator and load resistance values can be determined which lead to instability; they are entered on the Smith diagram as stability circles.

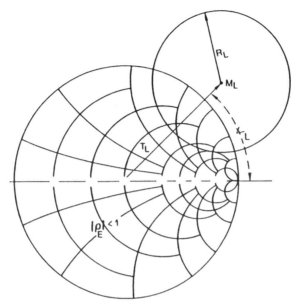

Figure 5.47   *Stability circle for output load* R_{CE} *of power transistor in Smith diagram. Impedance values within common area lead to instability*

Figure 5.47 shows, for example, a stability circle for the load resistance R_L. The transistor output may not be connected to the complex impedances which are included by the radius R_L in the Smith diagram.

   In addition to the nonlinear diode characteristics in the transistor, the voltage-dependent collector base capacitance C_{ob} in particular is the largest effective nonlinearity. The effect becomes practically input-dependent, which can lead to what is known as the parametric effect. The principle is, in short: if a capacitance diode is connected to a conventional lossy circuit in parallel, and pumped at double resonance frequency, an oscillator can result. A similar thing happens by the actuation (pumping) of C_{ob} by the working frequency: under 'favourable' impedance conditions of Z_{CE} for, for example, half, or possibly 1/3 or 1/4, of the operation frequency, this can be excited. A whole latticework of undesired signals can occur along with the working frequency! Remedies that have proved worthwhile are: a parallel capacitance C_p collector–emitter, wideband matching networks which provide real loads for the collector and base even outside the frequency band of interest, low-ohmic voltage supplies (good blocking), additional attenuation of approximately 1 dB and as a result an amount of decoupling in the intermediate matching networks, absorber filters in harmonic filter technology, and, if necessary, circulators. In most cases, the gain of a transistor specified by the manufacturer is not fully utilised in a cascade arrangement. It is, however, better to have only 23 dB amplification instead of 25 dB rather than an unstable amplifier, which at lower temperatures and conditions will become a power oscillator. A practical example is

Huth–Kühn oscillations at sufficiently large S_{12} by means of 'appropriate' base and collector reactances.

In view of the fact that S_{21} (forward gain of the transistor) becomes greater as frequency rises (approximately 6 dB/octave), one must take care to ensure that the values in the base and collector circuit are not too great at frequencies below the working range. R/C networks are used for this purpose, as shown in Figure 5.28: L_3 and C_1 are dimensioned in such a way that power at the working frequency cannot flow off to the voltage supply. At low frequencies, the output networks N_3 and N_4 no longer offer any effective real part. To prevent the transistor from then running high-ohmic under no load ($U_C > BV_{CB}$!), R_2 and R_3 function as ohmic equivalent loads; this means that the transistor is, in real terms, wideband-loaded. Similarly for the base circuit L_1, Fe, R_1.

5.3.4 Production of high HF power ratings

It is virtually impossible to connect power transistors in a direct parallel circuit for two reasons:

• Low input and output resistances make matching difficult, especially in wideband application.

• In a direct parallel connection, the inequality of the individual semiconductors and of the circuit cause mutual reciprocal effects (coupling → feedback), which leads to instability and a drop in amplification.

Accordingly, a suitable circuit must be provided which will carry out the parallel operation without reciprocal effects. The inputs and outputs of the individual amplifier modules must therefore be prevented from seeing each other; a decoupling effect of 10–20 dB is to be aimed for.

Figure 5.48 shows the solution of the problem. Two or more individual amplifiers V_1, V_2 (input and output impedance 50 Ω) are connected together with the help of what is known as a load distributor (LT)/load adder (LA). As a LT or LA, one can use hybrids (bridge transformers), Wilkinson arrangements, and a 3 dB 90° directional coupler.

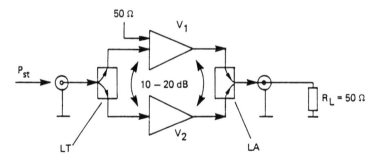

Figure 5.48 *Parallel connection of individual transistors*

5.3.4.1 Hybrids

The function of the hybrid LT or LA is described in Section 8.4.5. Figures 5.49a, b, c are intended to illustrate once again the main types used in practice for easier comprehension.

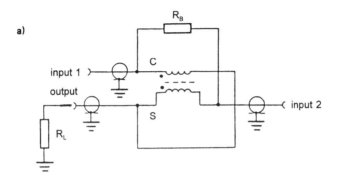

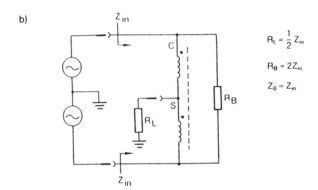

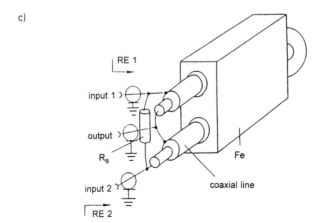

Figure 5.49 *Circuits for wideband power distribution or addition (1–500 MHz) with coaxial lines and ferrites (hybrids)*

As with line transformers, the considerations also apply in this case that, with large bandwidths, the length must be $L > \lambda/8$, and the line must be formed with ferrite (Fe). Depending on the mechanical size (maximum transferrable power), bandwidths of, for example, 1–500 (1000) MHz are possible (P = 500 W). A disadvantage is the occurrence of the 25 Ω impedance, e.g. at LA. To revert once again to 50 Ω, a transformer must be wired in a cascade with a resistance transformation ratio of 2:1. A conventional transformer cannot be manufactured for a large frequency range, so we must look for a power transformer which possesses these transformation properties at least in approximation. By applying the appropriate circuit arrangement, a resistance ratio of 9:4 can be created from three simple 1:1 Guanella transformers. In other words, 25 Ω instead of 50 Ω is transformed to $25 \times 9/4 = 56$ Ω. The VSWR is then 56 Ω/50 Ω = 1.12 which is a value entirely adequate in practice. The 25 Ω problem resolves itself if, for example, 4×50 Ω individual amplifiers have to be connected together because of the high summed power.

5.3.4.2 Wilkinson coupler

The basic principle of the simple Wilkinson coupler is described in Section 8.4.4, and for this reason only designs for large bandwidths are considered here. In view of the fact that this kind of power coupler is readily created in stripline technology (straightforward planar technology), it is frequently used in practice.

It is well known that wideband transformations can be carried out by means of cascaded lines of different wave resistances. The individual line lengths in this context feature a length predominantly of $\lambda/4$ in the mean transmission frequency. Using staggered lines of this type in the individual branches of a Wilkinson coupler, one can obtain transformation bandwidths of up to 10:1. It is also possible to set up couplers for N = 3:1, 4:1, 5:1 etc., in which the bandwidth naturally decreases if the number of elements is not increased accordingly.

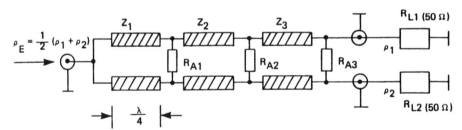

Figure 5.50   *Wilkinson scaler 2 $\leftrightarrow$ 1 for large bandwidths (100–500 MHz) in planar technology (N=2:1)*

5.3.4.3 3 dB 90° directional couplers

With the right dimensioning of the line couplers illustrated in Section 3.2.3.2, they can be used as LT or LA with special properties. A power P is fed in at the port 1, and half of it is offered to ports 2 and 4. The two output voltages are displaced to one another by 90° in the process. Connecting two individual amplifiers together with the aid of 90° 3 dB directional couplers (Figure 5.52), some interesting results occur considering the equations in Figure 5.51.

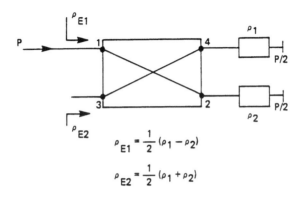

$$P_{E1} = \frac{1}{2}(P_1 - P_2)$$

$$P_{E2} = \frac{1}{2}(P_1 + P_2)$$

Figure 5.51   *Principle of 3 dB directional coupler*

The matching networks at the input of a power transistor are always measured in such a way that the best matching takes place at the highest frequency. The reflectance will therefore increase towards low frequencies, which means that, with constant forward power, a decreasingly effective control power will be attained, and thus a compensation of the transistor frequency response.

If we now switch amplifier inputs of this nature to a directional coupler 2, at outputs 2 and 4, the reflected power at these amplifier inputs will appear at port 3, and be converted into heat. Correspondingly, no reflected power occurs at port 1, and it therefore remains what may be termed a '50 Ω input'.

For the power addition, the two equal output powers, phase difference 90°, of the individual amplifiers are fed in at ports 3 and 4. The total power then appears at port 1. If the power values are not equal because of unequal amplification, or differ in phase, the differential amount appears at port 2. The outputs of the amplifiers are decoupled from one another (approx. 15–25 dB).

Figure 5.52 accordingly represents a 50 Ω module, which can now be used in further connection technology (Wilkinson scalers) as a basic module to form amplification units of any desired size.

Four individual modules according to Figure 5.52 are combined in the air navigation transmitter VD 490 from Rohde & Schwarz, with the aid of 4 ↔ 1 Wilkinson scalers to form the power amplifiers I and II (Figure 5.53). Their output power values are in turn connected together via 2 ↔ 1 Wilkinson scalers; the output power amounts to approximately 1200 W. The control power of about 30 W is divided in WT 1 onto two amplifier modules as shown in Figure 5.52. The output power of about 200 W receives WT 2 and divides it onto the inputs of the power amplifiers I and II.

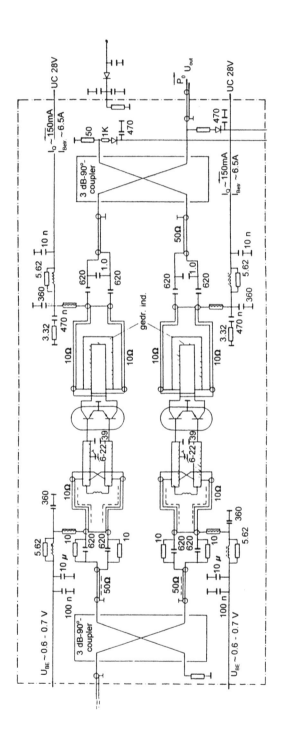

Figure 5.52 *UHF wideband amplifier $P_O = 200$ W, $G = 8.5$ dB, $= 200–430$ MHz*

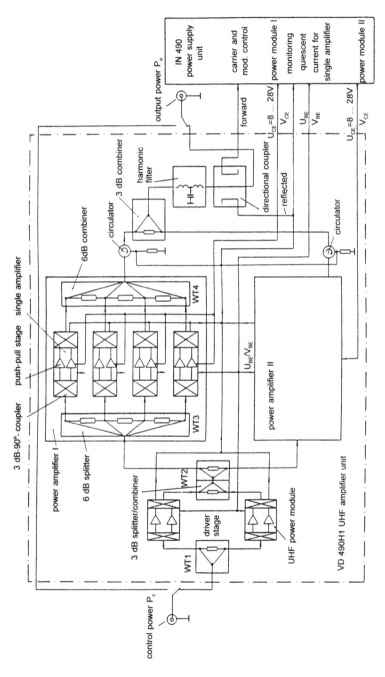

Figure 5.53   *Power amplifier for air navigation band 225–400 MHz, $P_{omax} = 1000\ W$*

Chapter 6

CAD, noise and optimisation

N. Krausse

6.1 Introduction

The computer has today won a secure place in solving problems in the technical sector. Computer-supported development work is described in general as CAE (computer aided engineering).

The development of HF circuits is an area well-suited for the application of CAE, since with the aid of the computer the electrical behaviour of even complex circuits can be calculated and optimised. In this Section the basic principles for HFCAE are dealt with. By this, is meant the analysis and optimisation of HF networks: program parts which go beyond this, such as automatic layout generation, are not dealt with here.

The problems discussed may be of a theoretical mathematical nature, but they are nevertheless closely linked to actual practice. In this context, we should remind ourselves of (vector) network analysers and noise measurement systems, by means of which it is possible, automatically (i.e. with control of the measurement point and error correction by computer) to

— measure the parameters required for CAE of many components (e.g. transistors)

— dimension the finished modules (e.g. amplifiers) in such a way that, in the final analysis, their measured behaviour matches that which has been calculated.

The intention of this section is to

— provide an insight into the steps for the analysis of HF circuits

— put readers in a position in which they can produce an efficient program system for themselves.

The following areas are considered, in three sections:

• Network analysis

• Noise analysis

• Network/noise optimisation

6.2 Network analysis

6.2.1 General

Experience has shown that network analysis is the most readily accessible of the three areas of network analysis, noise analysis and optimisation, and in general is also the most important because it enables rapid calculation to be made of the frequency and time behaviour of even complex circuits. Network analysis is based on the S-parameter theory. The following restrictions should apply to the considerations presented here:

- All networks are considered as two-port networks (also known as quadripoles, Figure 6.1). A two-port network is fully described by its S-parameters (and the reference wave impedance associated with them).

- All networks consist of at least one two-port network (Figure 6.2). A complex network can, in other words, be broken down into two-port networks that are connected to one another in a specific form.

- All networks are considered as linear, i.e. configurations which are inherently nonlinear (e.g. diodes, transistors) have such small signals imposed on them that they function as if they were linear (referred to as small-signal operation).

The following pages consider what connections are used to make up a larger network out of two-port networks.

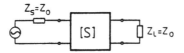

Figure 6.1   *Two-port network (quadripole), characterised by its S-parameter matrix*

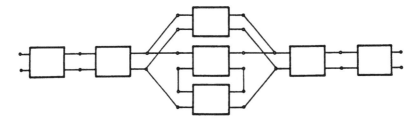

Figure 6.2   *Arrangement of two-port networks*

6.2.2 Ladder networks

The arrangement encountered most frequently is the ladder network (Figure 6.3). The formula required for the calculation is shown in Figure 6.3.

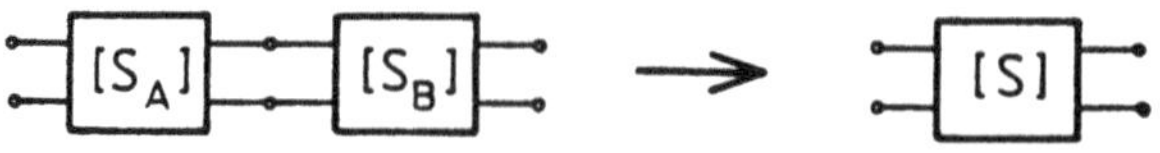

$$[S] = \begin{bmatrix} S_{11A} + \dfrac{S_{21A}\, S_{11B}\, S_{12A}}{1 - S_{11B}\, S_{22A}} & \dfrac{S_{12A}\, S_{12B}}{1 - S_{11B}\, S_{22A}} \\[3mm] \dfrac{S_{21A}\, S_{21B}}{1 - S_{11B}\, S_{22A}} & S_{22B} + \dfrac{S_{12B}\, S_{22A}\, S_{21B}}{1 - S_{11B}\, S_{22A}} \end{bmatrix}$$

Figure 6.3   *Ladder network of two-port networks*

6.2.3 Series connection

The series connection (e.g. a transistor with emitter impedance) is shown in Figure 6.4. The S-parameters of the two two-port networks that are to be connected must first be converted into Z-parameters. The equations required to do this are given in Chapter 1. The Z-parameters for series connections are obtained by the adding together the Z-parameters of the two two-port networks. The S-parameters of the connection are then obtained by converting on the basis of the Z-parameters, using the appropriate equations.

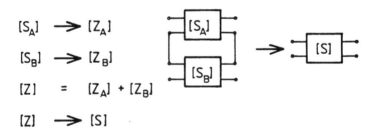

Figure 6.4   *Series connection of two-port networks*

6.2.4 Parallel connection

Figure 6.5 shows the parallel connection. In this context, first convert the S-parameters into Y-parameters (Chapter 1). The Y-parameters of the parallel connection are obtained by the addition of the Y-parameters of the two two-port networks. The Y-parameters are then converted back into S-parameters (Chapter 1).

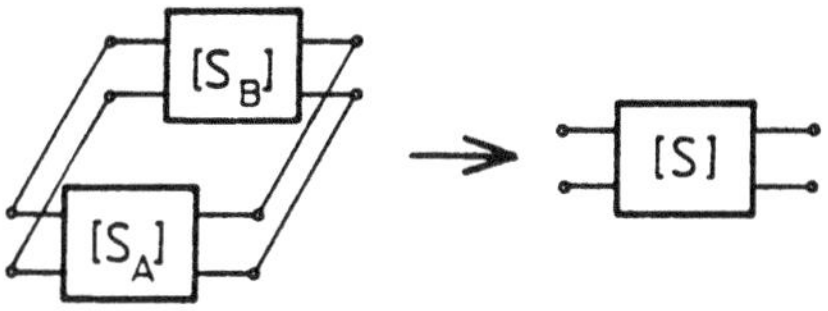

$$[S_A] \rightarrow [Y_A]$$

$$[S_B] \rightarrow [Y_B]$$

$$[Y] = [Y_A] + [Y_B]$$

$$[Y] \rightarrow [S]$$

Figure 6.5 *Parallel connection of two-port networks*

6.2.5 Changing the reference wave impedance

Assume, for example, that the S-parameters of a transistor are to be measured in a 50 Ω arrangement. The circuit into which the transistors are to be incorporated have, for some reason, a different impedance. Figure 6.6 shows the equations for the conversion.

$$S'_{11} = \frac{(S_{11}-\Gamma)(1-\Gamma S_{22})+\Gamma S_{12}S_{21}}{(1-\Gamma S_{11})(1-\Gamma S_{22})-\Gamma^2 S_{12}S_{21}}$$

$$S'_{21} = \frac{S_{21}(1-\Gamma^2)}{(1-\Gamma S_{11})(1-\Gamma S_{22})-\Gamma^2 S_{12}S_{21}}$$

$$S'_{22} = \frac{(S_{22}-\Gamma)(1-\Gamma S_{11})+\Gamma S_{12}S_{21}}{(1-\Gamma S_{11})(1-\Gamma S_{22})-\Gamma^2 S_{12}S_{21}}$$

$$S'_{12} = \frac{S_{12}(1-\Gamma^2)}{(1-\Gamma S_{11})(1-\Gamma S_{22})-\Gamma^2 S_{12}S_{21}}$$

$$\Gamma = \frac{Z'_0 - Z_0}{Z'_0 + Z_0}$$

Z_0: old reference wave impedance
Z'_0: new reference wave impedance

Figure 6.6 *S-parameters with changed reference wave impedance*

6.2.6 S-parameters of components

6.2.6.1 General

In view of the fact that the use of S-parameter theory is, in general, restricted to high and extra-high frequency engineering, only components which are used in these frequency ranges should be considered as elementary two-port networks (Figure 6.7).

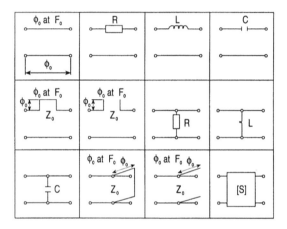

Figure 6.7 *Elementary modules in the network calculation*

There are four different basic types of component:

- line
- series or longitudinal impedance
- shunt admittance
- modules characterised by S-parameters

There are two principal ways of determining the S-parameters of a component:

— by calculation, if a component is correctly described in practical terms by one or more elementary modules

— by measurement, as in the case of transistors; passive components too, however, are sometimes better described by a measured S-parameter field.

S-parameters are measured with the help of suitable directional coupler arrangements (vector network analysers), which (with control and error correction by computer), supply the S-parameters very precisely over large frequency ranges, such as up to 40 GHz.

6.2.6.2 S-parameters of line section

Figure 6.8 shows a section of line with the wave impedance Z'_0, embedded in a system with the impedance Z_0.

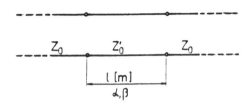

$$\Gamma = \frac{Z_0 - Z'_0}{Z_0 + Z'_0}$$

Figure 6.8   *Line section with changed wave impedance*

In addition to the wave impedance Z'_0 (which it is sensible to assume is a real value) the line is characterised by the following values:

- attenuation constant α [Np/m]

- phase constant $\beta = \dfrac{2\pi}{\lambda}$ [1/m] with λ = line wavelength

- physical length l [m]

Because the dimension [dB/m] is normally given, bear in mind that

$$\alpha \text{ [Np/m]} = 0.1151 \cdot \alpha \text{ [dB/m]}$$

It accordingly follows, from the formulas given in Figure 6.6, that for the S-parameters

$$p \quad = \exp\left[-\,l(\alpha + j\beta)\right]$$

$$S_{11} = S_{22} = \frac{\Gamma(p^2 - 1)}{1 - \Gamma^2 p^2}$$

$$S_{12} = S_{21} = \frac{(1 - \Gamma^2)p}{1 - \Gamma^2 p^2}$$

6.2.6.3 S-parameters of longitudinal impedance

Figure 6.9 shows the S-parameters of a two-port network, formed from a longitudinal impedance. This longitudinal impedance can be, as Figure 6.7 shows, a resistance, an inductance, a capacitance, or an open-circuited or short-circuited line.

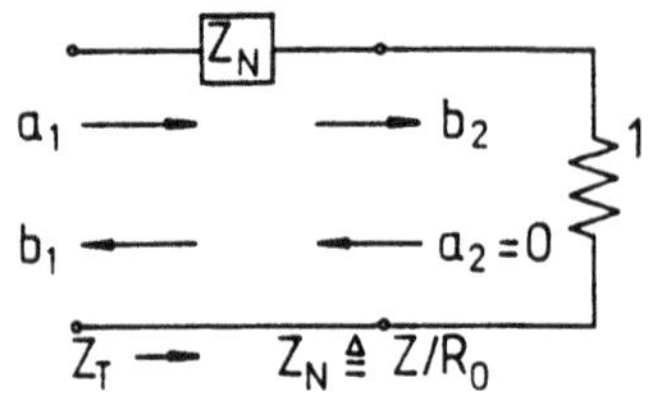

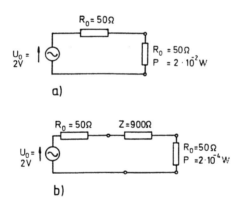

$$S_{11} = \frac{b_1}{a_1}\bigg|_{a_2=0} = \frac{Z_T - 1}{Z_T + 1}$$

$$S_{11} = \frac{Z_N + 1 - 1}{Z_N + 1 + 1} = \frac{Z_N}{2 + Z_N}$$

$$S_{21} = \frac{b_2}{a_1}\bigg|_{a_2=0} = \frac{a_1 - b_1}{a_1}$$

$$S_{21} = 1 - S_{11}$$

Figure 6.9   *S-parameters of longitudinal impedance*

An example is shown in Figure 6.10. A generator with the internal resistance $R_0 = 50\ \Omega$ and the no-load voltage $U_0 = 2$ V drives a load $R_0 = 50\ \Omega$. At R_0, the power is $P = 2 \cdot 10^{-2}$ W (Figure 6.10a).

Figure 6.10   *Example of longitudinal impedance*

(a) Matched
(b) Mismatched

In Figure 6.10b, a longitudinal impedance $Z = 900\ \Omega$ is inserted. From the formulas given in Figure 6.9, it follows that $Z_N = Z/R_0$ and therefore

$$S_{11} = 0.9$$
$$S_{21} = 0.1\ (\hat{=} - 20\ \text{dB})$$

The reflectance factor $S_{11} = 0.9$ has a positive qualifying symbol since the entire load impedance is greater than the internal impedance of the generator.

6.2.6.4 S-parameters of shunt admittance

Figure 6.11 shows the S-parameters of a two-pole network, formed from a shunt admittance. The same components come into question as with the longitudinal impedance.

Figure 6.12 shows an example analogous to Figure 6.10. In Figure 6.12a, for the sake of completeness, the match state is shown once again. In Figure 6.12b, a shunt admittance $Y = 0.36$ S is inserted.

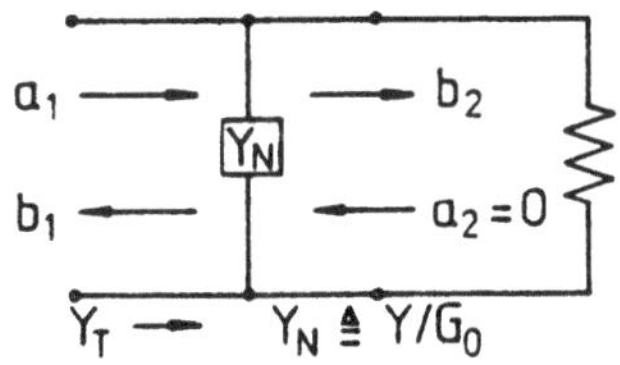

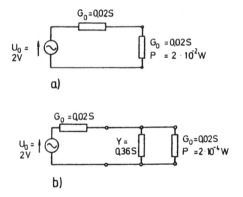

$$S_{11} = \frac{b_1}{a_1}\bigg|_{a_2=0} = \frac{1-Y_T}{1+Y_T}$$

$$S_{11} = \frac{1-1-Y_N}{1+1+Y_N} = -\frac{Y_N}{2+Y_N}$$

$$S_{21} = \frac{b_2}{a_1}\bigg|_{a_2=0} = \frac{a_1+b_i}{a_1}$$

$$S_{21} = 1 + S_{11}$$

Figure 6.11   *S-parameters of shunt admittance*

Figure 6.12   *Example of shunt admittance*

(a) Matched
(b) Mismatched

From the formulas given in Figure 6.11 it follows that $Y_N = Y/G_0 = 18$ and therefore

$$S_{11} = -0.9$$
$$S_{21} = 0.1 \ (\hat{=} -20 \text{ dB})$$

The reflectance factor S_{11} now has a negative qualifying symbol since the load impedance as a whole is smaller than the internal impedance of the generator.

6.2.7 Example: single-tuned bandpass filter

The task is to calculate a single-tuned bandpass filter with a mean frequency $f_0 = 1000$ MHz and a reverse attenuation of > 7 dB at $f_0 \pm 25$ MHz. An arrangement was selected which is formed of two shunt inductances connected by a length of line (Figure 6.13).

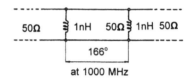

Figure 6.13   *Single-tuned bandpass filter*

The insertion loss is of special interest in considering the line losses. The following are used as HF cable:

> semirigid cable 0.141" with $Z = 50\ \Omega$
> dielectric Teflon ($\varepsilon = 2$)
> attenuation 0.43 dB/m at 1000 MHz

It can be seen that the reverse attenuation required is attained by inductances $L = 1$ nH, connected by a 50 Ω line of electrical length 166° at 1000 MHz. The electrical length 166°, at 1000 MHz and $\varepsilon = 2$, corresponds to a physical length of $l = 97.8$ mm.

To be able to calculate the ladder network from Figure 6.13, one needs the S-parameters of the shunt inductance from Figure 6.11, and those of the line from Figure 6.8. By application of the formulas in Figure 6.3, one can then finally determine the S-parameters of the bandpass filter.

Initially ignore the line losses and obtain the ideal curves from Figure 6.14a for $|S_{11}|$ and Figure 6.14b for $|S_{21}|$.

When taking into account the line losses of 0.43 dB/m, the insertion attenuation rises from 0 to 1.5 dB (Figure 6.15b), although the cable, about 100 mm long, has a pure line attenuation of only 0.04 dB. In addition to this, the reflectance factor deteriorates drastically (Figure 6.15a).

As well as examining the frequency response, a network analysis program also allows the determination of the time behaviour (time domain analysis). For this purpose, we break down a pulse, with the carrier function f_0, into its (sin x/x) spectrum, while restricting ourselves to a specific finite number of spectral lines above and below the carrier. The transmission behaviour of the individual lines is

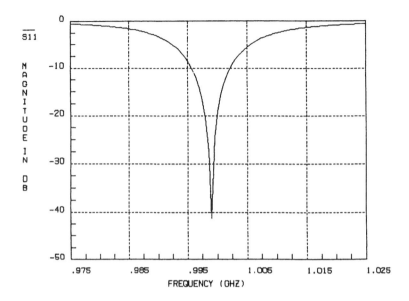

Figure 6.14a *Reflectance behaviour of loss-free bandpass filter*

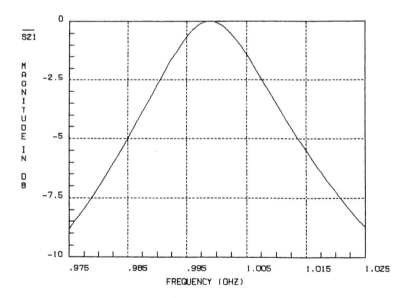

Figure 6.14b *Attenuation behaviour of loss-free bandpass filter*

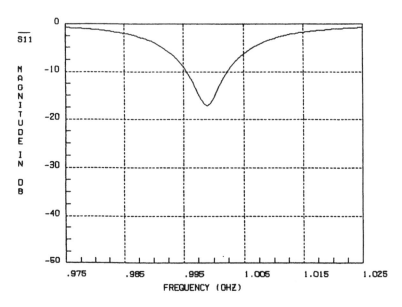

Figure 6.15a *Reflectance behaviour of lossy bandpass filter*

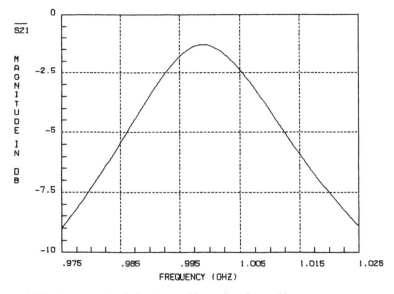

Figure 6.15b *Attenuation behaviour of lossy bandpass filter*

calculated with the aid of the analysis program; the lines are then added together again.

Example: Examine the time behaviour of a pulse after passage through the single-tuned bandpass filter, Figure 6.13. Let the pulse be characterised by

carrier frequency	1000 MHz
pulse interval	600 ns
pulse width	200 ns

The pulse is broken down into 255 spectral lines (in other words, 127 each to the left and right of the carrier). When run through the filter, the envelope curve shown in Figure 6.16 is obtained. One can identify the exponential transient and decay behaviour of the pulse.

From Figure 6.15b, 3 dB BW = 22.7 MHz. The (loaded) Q value of the circuit is therefore

$$Q = \frac{f_0}{BW} = \frac{1000 \text{ MHz}}{22.7 \text{ MHz}} = 44$$

The time t for the decay of an oscillation to the value $1/e = 0.37$ is $t = Q/\pi f_0 = 14$ ns. This value can be read off in Figure 6.16.

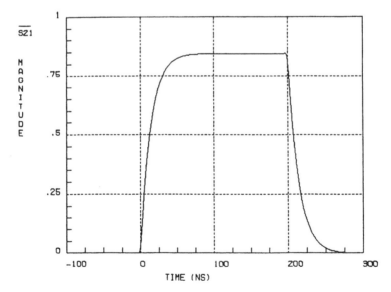

Figure 6.16 *Pulse after running through lossy bandpass, Figure 6.13*

6.3 Noise analysis

6.3.1 General

The fact that the noise coefficient plays a major part in the receiver input stage is well known. With radar receivers, HF preamplification has become established almost everywhere using GaAs FETs and HEMTs.

Up until a few years ago, it was impossible to calculate the noise coefficient of any random HF circuits. Pioneering work at the AEG Research Institute, however, finally provided a procedure for calculation which makes it possible to treat the problem of noise analysis in a similar manner to the S-parameter analysis.

6.2.3 Noise parameters of transistors

The values required for characterising the noise properties of a transistor are referred to as 'noise parameters', and consist of two real and one complex figure.

Consider Figure 6.17. A transistor contains a number of different noise sources in its interior. These can be replaced by an equivalent noise quadripole, connected in front of the transistor now assumed to be noise-free. The equivalent noise quadripole allows the signal to pass unhindered. This consists of a noise source in the series arm, and a noise source in the shunt arm.

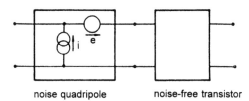

noise quadripole   noise-free transistor

Figure 6.17 *Noise sources from a transistor*

Noise sources i and e have, in part, the same causes, i.e. they are partially correlated with each other. Let the fraction of the noise current which is not correlated to the noise electromotive force be i_u, and let the other fraction, which is correlated with the noise voltage be i_c, so that

$$i = i_u + i_c \tag{6.1}$$

i_c is connected to e by means of the fictitious correlation conductance value Y_c

$$i_c = e \cdot Y_c \tag{6.2}$$

To be able to think of noise currents and voltages in terms of conductance values and resistances, one can use the definition

$$|\bar{e}|^2 = 4\,kT_0\,B\,R_n \tag{6.3}$$

$$|\overline{iu}|^2 = 4\,kT_0\,B\,G_u \tag{6.4}$$

where B = noise bandwidth
R_n = equivalent transistor noise impedance
G_u = noncorrelated transistor noise conductance value

The equivalent noise quadripole is fully described by the values R_n, G_u, and Y_C. These values are, however, of little use for a practical evaluation.

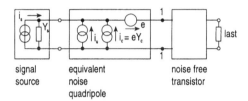

signal   equivalent   noise free
source   noise   transistor
                quadripole

Figure 6.18   *Noisy transistor with signal source*

Figure 6.18 shows a transistor with signal source and load. Assume the signal source to be delivering a noise current

$$|i_s|^2 = 4\,kT_0\,B\,G_s \tag{6.5}$$

To determine the total noise current, imagine the terminals 1-1 (Figure 6.18) to be short-circuited:

$$i_{tot} = i_s + i_u + i_c + e\,Y_s \tag{6.6}$$

and because $i_c = e\,Y_c$,

$$i_{tot} = i_s + i_u + e\,(Y_s + Y_c) \tag{6.7}$$

In view of the fact that the values which appear in eqn. 6.7 are not correlated with one another, the mean value square of i_{tot} is equal to the sum of the individual value squares:

$$|i_{tot}|^2 = |i_s|^2 + |i_u|^2 + |e|^2\,|Y_s + Y_c|^2 \tag{6.8}$$

The noise coefficient F can be defined by

$$F = \frac{\text{available total noise input power}}{\text{available noise power of the signal source}}$$

$$F = \frac{N_{tot}}{N_s} \tag{6.9}$$

Now,

$$N_{tot} = \frac{|\overline{i_{tot}}|^2}{4\,G_s} \tag{6.10}$$

and

$$N_s = \frac{|\overline{i_s}|^2}{4\,G_s} \tag{6.11}$$

and therefore

$$F = \frac{|\overline{i_{tot}}|^2}{|\overline{i_s}|^2} \tag{6.12}$$

and with eqn. 6.8,

$$F = 1 + \frac{|\overline{i_u}|^2 + |\overline{e}|^2\,|Y_c + Y_s|^2}{|\overline{i_s}|^2} \tag{6.13}$$

With eqns. 6.3 – 6.5, it follows that

$$F = 1 + \frac{G_u}{G_s} + \frac{R_n}{G_s}\,|Y_s + Y_c|^2 \tag{6.14}$$

and with

$$Y_c = G_c + jB_c$$

$$F = 1 + \frac{G_u}{G_s} + \frac{R_n}{G_s}\,[\,(G_s + G_c)^2 + (B_s + B_c)^2\,] \tag{6.15}$$

According to eqn. 6.15, the noise coefficient F is a function of the source admittance $Y_S = G_S + jB_S$. The question now arises of what value must be allotted to G_S and B_S to obtain the minimum noise coefficient F_{min}. From eqn. 6.15 it can immediately be seen that $B_S + B_C = 0$, since F is a minimum, i.e.

$$B_0 = B_S = -B_C \text{ for } F_{min} \tag{6.16}$$

B_0 is therefore the source susceptance for the minimum noise coefficient. To find the optimum G_S, differentiate eqn. 6.15 according to G_S. This leads to

$$G_0 = G_S = \sqrt{\frac{G_u + R_n G_c^2}{R_n}} \text{ for } F_{min} \tag{6.17}$$

G_0 is therefore the source conductance for the minimum noise coefficient. Equations 6.16 and 6.17 are inserted in eqn. 6.15:

$$F_{min} = 1 + 2 R_n (G_0 + G_c) \tag{6.18}$$

Extracting $B_0 = -B_C$ from eqn. 6.17

$$G_u = R_n (G_0^2 - G_c^2)$$

and introducing eqn. 6.18 into eqn. 6.15, it follows that

$$F = F_{min} + \frac{R_n}{G_S} |Y_S - Y_0|^2 \tag{6.19}$$

or

$$F = F_{min} + \frac{R_n}{G_S} [(G_S - G_0)^2 + (B_S - B_0)^2] \tag{6.20}$$

This expression for the noise coefficient contains the values F_{min}, R_n required for characterising the noise properties of a transistor, as well as the complex Y_0.

It is desirable to obtain the smallest possible noise coefficient of a two-port network. F_{min} is attained when we offer the source admittance Y_0 to the transistor. This is what is known as 'noise matching', which is normally not identical to power matching.

A brief word about measuring the noise parameters: eqn. 6.20 contains four unknowns; by measuring the noise coefficient at four different source reflectance factors, four equations can be produced with four unknowns. Normally, 6–8 measurements are made, since measurement errors have a relatively strong effect. Many manufacturers indicate the noise parameters of a transistor (generally in tabular form), which naturally shortens development work considerably.

6.3.3 Noisy networks

It is not possible to make use of the noise parameters immediately to calculate the noise properties of any particular random network. Rather, a matrix similar to one of the S-parameters initially needs to be created. Figure 6.19 shows what is referred to as the E-matrix for quadripoles that are to be characterised by noise parameters. Longitudinal admittances are described by a C-matrix (Figure 6.20), and shunt impedances by a D-matrix (Figure 6.21). The following applies to the conversion of the C-matrix into the E-matrix:

$$E = N^{-1} \cdot C \cdot N^{-1+} \tag{6.21}$$

where

$$N^{-1} = \begin{bmatrix} 0 & A_{12} \\ 1 & A_{22} \end{bmatrix} \tag{6.22}$$

(For the calculation of the A-parameters from the S-parameters, see Chapter 1).

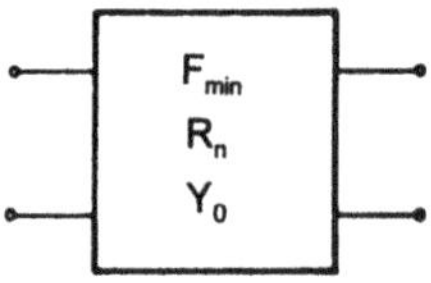

$$[E] = \begin{bmatrix} R_n & ((F_{min}-1)/2)-R_nY_0^* \\ ((F_{min}-1)/2)-R_nY_0 & R_n|Y_0|^2 \end{bmatrix}$$

Figure 6.19   *E-matrix of a transistor*

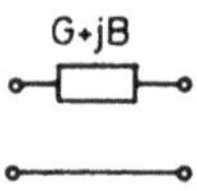

$$[C] = \begin{bmatrix} G & -G \\ -G & G \end{bmatrix}$$

Figure 6.20   *C-matrix of longitudinal admittance*

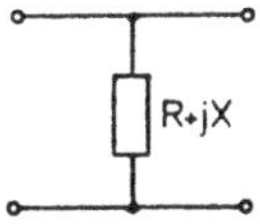

$$[D] = \begin{bmatrix} R & R \\ R & R \end{bmatrix}$$

Figure 6.21 *D-matrix of a shunt impedance*

In eqn. 6.21, N^{-1+} is the Hermitian conjugate of N^{-1}. Let

$$X = \begin{bmatrix} X_{11} & X_{12} \\ X_{21} & X_{22} \end{bmatrix} \qquad (6.23)$$

The Hermitian conjugated form is then

$$X^{+} = \begin{bmatrix} X^{*}_{11} & X^{*}_{22} \\ X^{*}_{12} & X^{*}_{22} \end{bmatrix} \qquad (6.24)$$

For the calculation of the C-matrix from the E-matrix,

$$C = N \cdot E \cdot N^{+} \qquad (6.25)$$

where

$$N = \begin{bmatrix} -Y_{11} & 1 \\ -Y_{12} & 0 \end{bmatrix} \qquad (6.26)$$

Further, the E-matrix can be calculated from the D-matrix:

$$E = M^{-1} \cdot D \cdot M^{-1+} \qquad (6.27)$$

where

$$M^{-1} = \begin{bmatrix} 1 & -A_{11} \\ 0 & -A_{21} \end{bmatrix} \qquad (6.28)$$

The conversion of the E-matrix into the D-matrix is

$$C = M \cdot E \cdot M^{+} \qquad (6.29)$$

where

$$M = \begin{bmatrix} 1 & -Z_{11} \\ 0 & -Z_{21} \end{bmatrix}$$ (6.30)

One also needs the C- and D-matrices to calculate the noise properties of series and parallel circuits (Figure 6.22).

Type	Optimisation	Noise
Series	$[Z] = [Z_1] + [Z_2]$	$[D] = [D_1] + [D_2]$
Parallel	$[Y] = [Y_1] + [Y_2]$	$[C] = [C_1] + [C_2]$
Ladder	$[A] = [A_1] \cdot [A_2]$	$[E] = [A_1][E_2][A_1]^+ + [E_1]$

Figure 6.22 *Interconnection of elementary two-port networks*

When the E-matrix of a network is finally calculated, it follows, for the noise coefficient, that

$$F = \text{Re}(E_{11}) + \text{Re}(E_{12}) + \text{Re}(E_{21}) + \text{Re}(E_{22})$$ (6.31)

6.4 Network optimisation

6.4.1 General

The term 'optimisation' (of a circuit) can be defined as follows:

> In a circuit with a specific number of parameters, these parameters should be adjusted in such a way that the behaviour of the circuit satisfies requirements previously defined.

What this means is that a circuit is initially designed in a conventional manner. During the subsequent analysis, it is generally discovered that the circuit behaviour required (e.g. wideband amplification) will not be attained. When carrying out optimisation with the aid of the computer, an efficient mathematical procedure provides for a variation in the component parameter in such a way that the desired circuit process is achieved within the framework of the possibilities provided by the circuit.

An example may illustrate this (Figure 6.23). A transistor characterised by an S-parameter matrix is connected to an input and output port with matching networks. The behaviour of the networks is shown by the dotted curves. What is required, however, is the smallest possible reflectance factors $|S_{11}|$ and $|S_{22}|$, as well as amplification of 12.5 dB. The computer is provided with six parameters for variation. The results are shown by the unbroken curves: a substantial improvement. One can see how the component parameters have been altered.

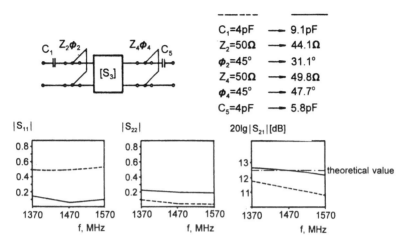

Figure 6.23 *Example of circuit optimisation*

More substantial improvement (in particular of $|S_{22}|$) is not possible because of the effect of the reactive elements exerting a negative effect on the other values.

What is in fact the task of optimisation? Consider, for example, the amplification $20 \log |S_{21}|$ in Figure 6.23. The intention is that at each of the three frequency points the difference between actual value and reference value should be as small as possible or, to formulate this correctly in mathematical terms, that the sum of the squares of the differences at the support points should approach zero (square to eliminate the effect of sign differences and to emphasise larger differences):

$$\text{error function } FF = \sum_{F_1}^{F_n} (G_{actual} - G_{ref})^2 \to 0$$

Other values which need to be optimised can also be adopted into the optimisation or error function, such as $|S_{11}|$, $|S_{22}|$ and the noise coefficient. In this context, different reference values with different weightings can be selected at the individual frequencies:

$$FF = \sum_{F(x=1)}^{F(x=n)} WG_x (G_{actual,x} - G_{ref,x})^2$$

$$+ WS_{11x} (|S_{11}|_{actual,x} - |S_{11}|_{ref.x})^2 + \ldots \qquad (6.32)$$

In this formula

$F(x=1)$, $F(x=n)$: frequencies $F_1...F_n$
WG_x : weighting for amplification of the frequency F_x
WS_{11x} : weighting for $|S_{11}|$ (etc.) at the frequency F_x.

Essentially, the mathematical task is to search for the minimum of a function of many variables (six variables in the example in Figure 6.23). To do this, mathematical processes have been developed, one of which is presented in the following Section.

6.4.2 Optimisation according to Fletcher and Reeves

In 1964, R. Fletcher and C.M. Reeves, in their paper entitled 'Function minimisation by conjugate gradients', described an efficient method of searching for the local minimum of a function of several variables. The corresponding FORTRAN program is listed below, together with an example. It consists of the following parts:

(1)   SUBROUTINE FKT (X,Y)
      This is where the function value to be minimised $Y(X(1)...X(N)$ is determined.

(2)   SUBROUTINE QUADIN
      Calculation of the coefficients A0, A1, and A2 of the equation

$$Y = A0 + A1 \cdot x + A2 \cdot x^2$$

      from three value pairs x0|y0, x1|y1 and x2|y2.

(3)   SUBROUTINE FUNCT(N, X0, F, G)
      Calculation of the gradient G of FKT(X0,F) with N variables by quadratic interpolation with the values X0(1) to X0(N).

(4)(a) FUNCTION DOT(N,A,B)
    (b) FUNCTION UPDOT(N,A,B,I)
    (c) SUBROUTINE FLEREM
      FLEREM, together with DOT and UPDOT, is the actual minimisation program.

(5)   OPTTEST.FOR
      This is the main program, with input of the values still required, and the output of the result. The listing shows the program sequence in which the function to be minimised is defined in the first subroutine.

```
      SUBROUTINE FKT(X, Y)
      DIMENSION X(20)
C     Example:  On p. 158 of the paper by
C     M.J.D. Powell, "An efficient method for finding the minimum of
C     a function of several variables without calculating derivatives",
C     Computer Journal Vol. 7,
C     1964, p. 155 ... 162
      P=1.570796326
      Y=-1./(1.+(X(1)-X(2))**2)-SPIN(P*X(2)*X(3))-EXP(-((X(1)+(x(3))
     *   /X(2)-2.)**2)
```

```
      RETURN
      END

      SUBROUTINE QUADIN(X0,X1,X2,Y0,Y1,Y2,A0,A1,A2)
C     Quadratic interpolation of the function y = A0 + A1*x + A2*x^2
C     Calculation of A0, A1, A2 from three value pairs
      G1X1-(Y1-Y0)/(X1-X0)
      G1X2-(Y2-Y0)/(X2-X0)
      A2=(G1X2-G1X1)/(X2-X1)
      A0=Y0-G1X1*X0+A2*X0*X1
      A1=G1X1-A2*(X0+X1)
      RETURN
      END

      SUBROUTINE FUNCT(N, X0, F, G)
C     Calculation of the gradient G of a function with the
C     values X0(1) to X0(N), and the formation of
C     derivation 1
      DIMENSION X0(20), G(20), X(20)
      DO 1 I=1,N
    1 X(I)=XO(I)
      CALL FKT(XO,F)
      DO 20 I=1,N
      IF (ABS(XO(I)) .LT. 0.0001) GOTO 10
      DX=0.001*XO(I)
      GOTO 11
   10 DX=0.0001
   11 X(I)=XO(I)-DX
      CALL FKT(X,YLI)
      XLI=X(I)
      X(I)=XO(I)+DX
      CALL FKT(X,YRE)
      XRE=X(I)
      CALL QUADIN(XLI,XO(I),XRE,YLI,F,YRE,AO,A1,A2)
      G(I)=A1+2.*A2*XO(I)
      X(I)=XO(I)
   20 CONTINUE
      RETURN
      END
      FUNCTION DOT(N,A,B)
      DIMENSION A(20), B(20)
      S=0.0
      DO 10 I=1,N
   10 S=S + A(I)*B(I)
      DOT=S
      RETURN
      END

      FUNCTION UPDOT(N,A,B,I)
      DIMENSION A(20), B(20)
      K=I
      S=0.0
      IM1=I-1
```

```
          DO 10 J=1,IM1
          S=S + A(K)*B(J)
          K=K+N-J
    10 CONTINUE
          DO 20 J=1,N
    20 S=S + A(K+J-I)*B(J)
          UPDOT=S
          RETURN
          END


          SUBROUTINE FLEREM(N,X,F,EST,EPS,CONV,LIMIT,H.LOADH)
C      R. Fletcher, C. M. Reeves, "Function minimization by conjugate
C      gradients", Computer Journal Vol. 7, 1964, p. 149 ... 154
          LOGICAL CONV, LOADH
          DIMENSION X(20),H(210),G(20),S(20),GAMMA(20),SIGMA(20)
C      Field size of H is N*(N+1)/2
          DO 10 I=1,N
    10 G(I)=0.0
C
C      SET INITIAL H
          IF (LOADH) THEN
          K=1
          D0 30 I=1,N
          H(K)=1.0
          NI=N-I
          DO 20 J=1,NI
    20 H(K+J)=0.0
          K=K+N-I+1
    30 CONTINUE
          END IF
C      END FORMATION OF UNIT MATRIX IN H
C
C      START OF MINIMISATION
          CONV=.TRUE
          CALL FUNCT(N,X,F,G)
          OLDF=F+ABS(0.001*F)
          DO 150 WHILE (OLDF .GT. F)
          DO 140 ICOUNT=1,ICOUNT+1
          OLDF=F
          DO 40 I=1,N
          SIGMA(I)=X(I)
          GAMMA(I)=G(I)
          S(I)=-UPDOT(N,H,G,I)
    40 CONTINUE
C      END PRESERVATION OF X, G AND FORMATION OF S
C
C      SEARCH ALONG S
          YB=F
          VB=DOT(N,G,S)
          SS=DOT(N,S,S)
          IF ((ICOUNT .EQ. 1) .AND. (SQRT(SS) .LT. EPS)) GOTO 999
          IF (VB .GE. EPS) GOTO 90
          AK=2.*(EST-F)/VB
```

```
      IF ((AK .GT. 0.) .AND. (AK*AK*SS .LT. 1.)) THEN
      AH=AK
      ELSE
      AH=1./SQRT(SS)
      END IF
      AK=0
   50 YA=YB
      VA=VB
      DO 60 I=1,N
   60 X(I)=X(I)+AH*S(I)
      CALL FUNCT(N,X,F,G)
      YB=F
      VB=DOT(N,G,S)
      IF ((VB .LT. 0.) .AND. (YB .LT. YA)) THEN
      AK=AH+AK
      AH=AK
      GOTO 50
      END IF
      T=0.
   65 Z=3.*(YA-YB)/AH+VA+VB
      W=SQRT(Z*Z-VA*VB)
      AK=AH*(VB+W-Z)/(VB-VA+2.*W)
      DO 70 I=1,N
   70 X(I)=X(I)+(T-AK)*S(I)
      CALL FUNCT(N,X,F,G)
      IF ((F .GT. YA) .OR. (F .GT. YB)) THEN
      VC=DOT(N,G,S)
      IF (VC .LT. 0.) THEN
      YA=F
      VA=VC
      AH=AK
      T=AH
      ELSE
      YB=F
      VB=VC
      T=O
      AH=AH-AK
      END IF
      GOTO 65
      END IF
C     END OF SEARCH ALONG 5
C
   90 DO 100 I=1,N
      SIGMA(I)=X(I)-SIGMA(I)
      GAMMA(I)=G(I)-GAMMA(I)R    100    CONTINUE
      SG=DOT(N,SIGMA,GAMMA)
      IF (INCOUNT .GE. N) THEN
      IF ((SQRT(DOT(N,S,S)) .LT. EPS) .OR. (SQRT(DOT(N,SIGMA,SIGMA))
     *.LT. EPS)) GOTO 999
      END IF
C     END TEST FOR VANISHING DERIVATIVE
C
      DO 110 I=1,N
```

```
100  S(I)=UPDOT(N,H,GAMMA,I)
     GHG=DOT(N,S,GAMMA)
     K=1
     DO 130 I=1,N
     DO 120 J=I,N
     H(K)=H(K)+SIGMA(I)*SIGMA(J)/SG - S(I)*S(J)/GHG
     H=K+1
120  CONTINUE
130  CONTINUE
C    END UPDATING OF H
C
     IF (ICOUNT .GT. LIMIT) GOTO 998
140  END DO
150  END DO
C    END OF LOOP CONTROLLED BY ICOUNT
     GOTO 999
998  CONV=.FLASE.
999  RETURN
     END

C    FILE OPTTEST.FOR 11-MAI-89
     DIMENSION X(20), H(210)
     LOGICAL CONV
     DATA X/20*0.0/, h/210*0.0/, F/0.0/
     WRITE(6,10)
10   FORMAT(' Number N of the variables ? -- max. 20 permitted')
     READ(5,*) n
     WRITE(6,20)
20   FORMAT(' Precision EPS of the iteration ?')
     READ(5,*) EPS
     WRITE(6,30)
30   FORMAT(' Number LIMIT of the iterations ?')
     READ(5,*) LIMIT
     WRITE(6,40) N
40   FORMAT (1X, I2,' estimated values X, separated by commas, input')
     READ(5,*) (X(I), I=1,N)
     WRITE(6,45)
45   FORMAT(' estimated value EST of the function value at the minimum ?')
     READ(5,*) EST
     WRITE(6,50) EPS,LIMIT,EST
     WRITE(7,50) EPS,LIMIT,EST
50   FORMAT(///9X, ' input values: '/9X
    *'EPS   = ',FB.6/9X,
    *'LIMIT = ',I2/9X,
    *'EST   = ',E18.10/)
     DO 70 I=1,N
     WRITE(6,60) I,X(I)
     WRITE(7,60) I,X(I)
60   FORMAT(9X,
    *'X(',12,') = ',E18. 10)
70   CONTINUE
     CALL FLEREM(N,X,F,EST,EPS,CONV,LIMIT,H,.TRUE.)
     WRITE(6,80)
```

```
      WRITE(7,80)
   80 FORMAT(///9X, 'Solution: '/)
      DO 90 I=1,N
      WRITE(6,60) I,X(I)
      WRITE(7,60) I,X(I)
   90 CONTINUE
      WRITE(6,100) F
      WRITE(7,100) F
  100 FORMAT(//9X, 'function value: '/9X,
     *'F = ',E18.10)
      STOP
      END
```

```
Input values:
EPS   = 0.000100
LIMIT =  3
EST   =   0.1500000000E+01

X( 1) =   0.0000000000E+00
X( 2) =   0.1000000000E+01
X( 3) =   0.2000000000E+02
```

```
Solution:

X( 1) =   0.100028073BE+01
X( 2) =   0.1000211477E+01
X( 3) =   9.9998677969E+00
```

```
Function value:
F = -0.3000000000E+01
```

6.4.3 *Example of optimisation (T-attenuator)*

Figure 6.24 shows a symmetrical T-attenuator, formed from two series resistances X(1) and the shunt resistance X(2).

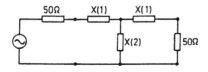

Figure 6.24   *T-attenuator*

It is required to determine X(1) and X(2) for a 50 Ω system in such a way that

$|S_{11}| = 0.0$ (matching)
$|S_{21}| = 0.5$ (attenuation 6 dB)

The first part of the corresponding FORTRAN program appears in the following listing.

(1) SUBROUTINE CSCY: Calculation of the S-parameters of a complex transverse conductance value CY (see Figure 6.11)

(2) SUBROUTINE CSCZ: Calculation of the S-parameters of a complex series resistance CZ (see Figure 6.9)

(3) SUBROUTINE PROFAN: Calculation of the ladder network from X(1), X(2), X(1) (see Figure 6.3)

(4) SUBROUTINE FKT: Input of the reference values and calculation of the error function.

The other part of the program is identical to that described in Section 6.4.2 (from SUBROUTINE QUADIN).

```
      SUBROUTINE CSCY(CS11,CS12,CS21,CS22,CY)
C     S-parameters of a complex shunt conductance CY
      COMPLEX CS11,CS12,CS21,CS22,CY
      CS11=-CY/C2.+CY)
      CS22=CS22
      CS12=1.-CS11
      CS21=CS12
      RETURN
      ENDR
      SUBROUTINE CSCZ(CS11,CS12,CS21,CS22,CZ)
C     S-parameters of a complex series resistance CZ
      COMPLEX CS11,CS12,CS21,CS22,CZ
      CS11=-CZ/C2.+CZ)
      CS22=CS22
      CS12=1.-CS11
      CS21=CS12
      RETURN
      END

      SUBROUTINE PROFAN(X)
C     Calculation of the S-parameters of the T-attenuator
      DIMENSION X(20),CS11(3),CS12(3),CS21(3),CS22(3)
      COMPLEX CS11,CS12,CS21,CS22,CS11N,CS12N,CS21N,CS22N,CZ,CY,CNEN
      COMMON /BLK1,CS11N,CS21N
      Z1=X(1)/50.
      CZ=CMPLX(Z1,0.)
C     S-parameters of the series resistor 1
      CALL CSCZ(CS11(1),CS12(1),CS21(1),CS22(1),CZ)
C     S-parameters of the series resistor 2
```

```
      CS11(3)=CS11(1)
      CS12(3)=CS12(1)
      CS21(3)=CS21(1)
      CS22(3)=CS22(1)
      Y1=50./X(2)
      CY=CMPLX(Y1,0.)
C     S-parameters of the shunt conductance value
      CALL CSCY(CS11(1),CS12(1),CS21(1),CS22(1),CY)
C     Ladder network
      DO 10 I=2,3
      IC=I-1
      CNEN=1.-CS11(I)*CS22(IC)
      CS11N=CS11(IC)+CS21(IC)*CS11(I)*CS12(IC)/CNEN
      CS12N=CS12(IC)*CS12(I)/CNEN
      CS21N=CS21(IC)*CS22(I)/CNEN
      CS22N=CS22(I)+CS12(I)*CS22(IC)*CS21(I)/CNEN
      CS11(I)=CS11N
      CS12(I)=CS12N
      CS21(I)=CS21N
      CS22(I)=CS22N
   10 CONTINUE
      RETURN
      END

      SUBROUTINE FKT(X,FF)
C     Calculation of the error function FF
      COMPLEXCS11N,CS21N
      DIMENSION X(20)
      COMMON /BLK1/CS11N,CS21N
      IAUF=IAUF+1
      CALL PROFAN(X)
      IF (IAUF .GT. 1) GOTO 40
      WRITE(6,10)
   10 FORMAT(//' Desired reflectance factor |S11REF| ?')
      READ(5,*) S11SOLL
      WRITE(6,20)
   20 FORMAT(//' Desired transmission factor |S21REF| ?')
      READ(5,*) S21SOLL
      WRITE(6,30) S11SOLL,S21SOLL
      WRITE(7,30) S11SOLL,S21SOLL
   30 FORMAT(/9X,
     *'|S11SOLL| = ',F5.3/9X,
     *'|S21SOLL| = ',F5.3)
C
C     Error function
   40 FF=(CABS(CS11N)-S11SOLL)**2 + (CABS(CS21N)-S21SOLL)**2
      RETURN
      END
```

The result for 1 Ω start values according to three iterations is shown: the correct resistance values are attained; the (error) function value is very small.

```
Input values:
EPS   = 0.000100
LIMIT =  3
EST   =   0.0000000000E+00

X( 1) =   0.1000000000E+01
X( 2) =   0.1000000000E+01

|S11REF| = 0.000
|S21REF| = 0.500


Solution:

X( 1) =   0.1666666794E+02
X( 2) =   0.6666670227E+02


Function value:
F =   0.2670086374E-13
```

Chapter 7

Survey of CAD programs

F. Schmehr

7.1 Introduction

Until a few years ago, high investment costs meant that computer aided design, or CAD for short, was almost exclusively the prerogative of large commercial concerns. Essentially, until then the term CAD had been understood to mean computer-aided development and design in, for example, the mechanical engineering sector or for resolving PCB tasks in electrical engineering. Thanks to the rapid pace of development in the software sector, a large number of specialised areas have since arisen for which the expression computer assisted engineering (CAE) can be applied as an umbrella term for all computer-assisted activities in the development sector.

A number of highly efficient programs have been developed for the analogue and digital high-frequency technology sectors, which make it possible to achieve the synthesis, analysis and optimisation of an electronic circuit, right through to the automatic generation of a layout.

7.2 Historic development of high-frequency engineering

1960–1970

– Microstrip technology
– Further development of filter synthesis
– S-parameters
– Microwave transistors
– Network analysers
– Timesharing operation of mainframe computers
– SPEEDY and GENERAL programs (L. Besser)

1970–1980

– GaAs technology
– MIC technology
– Synthesis of matching networks
– CAE programs COMPACT, FILSYN, MICROCOMPACT
   (L. Besser, G. Szentirmai, C. Holmes)
   AMPSYN, CADSYN, DEELAY

1980–present

– Series production of MIV circuits
– Reasonably priced, high-capacity computers
– SUPERCOMPACT, SUPERFILSYN, AUTOART

7.3 CAE programs for high-frequency development

The aim of this section is to introduce programs from Compact Software, Inc., and from Dr. G. Szentirmai. For reasons of space, only the COMPACT, SUPERCOMPACT, and SUPERFILSYN programs are introduced in detail.

7.3.1 COMPACT

This program provides analysis and optimisation of active and passive circuits in the MF, HF and microwave frequency range. It has distributed elements, an extensive semiconductor databank. It features sensitivity and MONTECARLO analysis, noise and group delay time analysis and optimisation, analysis and optimisation of circuits containing n-port networks (e.g. circulators, multiplexers, diplexers, etc.), cables, striplines or hollow conductors and many plot possibilities, including Smith diagrams.

7.3.2 MICROCOMPACT

This package essentially corresponds to that of the COMPACT program, but does not contain the program sections for calculating striplines/microstriplines and for tolerance analysis.

7.3.3 SUPERCOMPACT

Some major developments have been implemented in comparison with COMPACT: the synthesis of matching networks, substrate databank. extended tolerance analysis, noise parameter databank, all microchip elements including branches (max. four times), substantially extended graphics possibilities, processing of filter descriptions which were generated by SUPERFILSYN, and production of a circuit description for the generation of layouts together with the AUTOART program.

7.3.4 AUTOART

This generates two-dimensional layouts for microwave circuits and MICs. Automatic conversion of circuit models into layout sections. Interactive working on graphics terminal. Comprehensive component element databank. Generation of layouts for a circuit calculated with SUPERCOMPACT.

7.3.5 FILSYN and SUPERFILSYN (Dr. G. Szentirmai)

This is by far the most effective program for the synthesis and analysis of practically all conventional filter types. The transfer function is calculated from the filter specifications. Realisation can be achieved in the form of a passive LC, active RC, digital (IIR or FIR) or microwave filter, and in part with automatic determination of the filter structure or the minimal required word width in the case of digital filters.

7.3.6 AMPSYN, CADSYN and DEELAY

These programs provide the following features:

AMPSYN: Direct synthesis of matching networks formed of concentrated elements in the HF and microwave sector (for nonlinear power amplifiers also).

CADSYN: Direct synthesis of matching networks from striplines in microwave operation.

DEELAY: Dimensioning of allpass networks for group delay time compensation of filters and transfer systems, in which COMPACT can be subsequently used for optimisation.

7.4 Circuit analysis, synthesis and optimisation

It is necessary to make a clear distinction between the tasks of analysis, synthesis and optimisation.

7.4.1 Analysis

Analysis is the calculation of the response of a *known* system to a given form of excitation (e.g. step response). It follows that a correctly posed analysis problem *always* has a solution. The majority of circuit calculation programs are *pure* analysis programs.

7.4.2 Synthesis

Synthesis is a systematic process for discovering a circuit which satisfies the specifications. It is not certain whether a solution exists. Compared with trial and error or crude and approximate approaches, formal synthesis has many advantages including providing a direct way to the solution and having a low computer time requirement. However, it does have the disadvantages that no procedure exists for some tasks which are posed and that element values can only be kept under control with difficulty. In many cases the mathematical development is difficult.

The process of synthesis has a number of stages consisting of the following:

- Determination of specification

- Selection of circuit configuration

- Calculation of circuit characteristics

- Calculation of element values

- If element values are unfavourable or incapable of being realised:
  Repetition of the steps outlined above

- Analysis, setting up test, test.

7.4.3 Optimisation

For a given circuit with a particular number of parameters, these parameters should be adjusted in such a way that the circuit behaviour satisfies the requirements that have been specified beforehand. There are many practical task scenarios which cannot be resolved by synthesis alone and for which optimisation can provide a solution. Optimisation *must* be used if

- no suitable synthesis procedure exists

- the requirements are given as a curve or value table

- actual elements such as transistors, operation amplifiers, or imperfect reactive elements are used, and their actual properties have to be incorporated

– peripheral conditions regarding the circuit configuration or the element values are available.

Optimisation properties:

Mathematically, optimisation consists of minimisation of a scalar error function, which depends on variables (the optimisation parameters xl to xn). There are many procedures for this which exist in numerical mathematics. In general, no circuit topology change is permitted (i.e. no increase, for example, in the number of elements). The circuit which is to be optimised must be selected in such a way that it fulfils the requirements. If this is not possible, another configuration must be selected. It is not certain that the solution attained is always the best possible in each case.

Summary:

Optimisation is an iterative process in which continuous approximations are calculated in a desired characteristic, and which, if necessary, are improved until an adequate approximation is achieved. In view of the fact that this process is repeated very often, the number of iterations can be very large, and it is extremely time consuming. For circuits of comparable size, optimisation can require 10–100 times the amount of calculation required for synthesis. This means that optimisation should only be used if the task cannot be resolved by synthesis. Practical development tasks frequently require the careful mixing of synthesis and optimisation. The optimisation method resembles the trial and error method frequently applied in laboratory practice; in other words, the attempt is made, by more or less systematic change of the variable elements, to achieve the desired circuit behaviour. Optimisation requires refined mathematical methods, but in compensation can vary all the parameters simultaneously. On the other hand, empirical tuning for more than about five elements is practically impossible.

7.5 COMPACT program

7.5.1 Program structure

To facilitate entry into the above programs, an example using COMPACT is worked through first. The program structure is given in Figure 7.1.

7.5.2 Most important elements of circuit description

Examples of key elements in circuit description are as follows:

RES AA SE 10	Resistance AA, serial, 10 Ω
IND BB PA 15 (50 60)	Inductance BB, parallel, 15 nH $Q = 50$ at 60 MHz

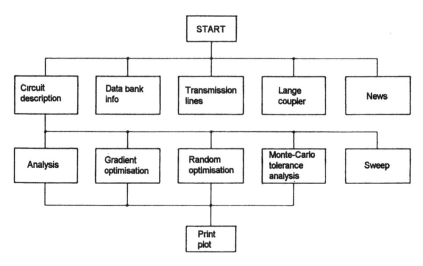

Figure 7.1   *Overall structure of COMPACT*

CAP CC SE 12 (500 30)	Capacitance CC, serial, 12 pF $Q = 500$ at 30 MHz
TWO DD S1 50	Active quadripole (transistor) Name DD data as S-parameter (S_1) in following databank, $Z_0 = 50\ \Omega$
SLC EE PA 380 56 197 25	Series circuit LC EE, parallel, 380 nH, 56 pF $Q = 197$ at 25 MHz.

7.5.3 Example of lowpass filter

(a) Analysis sequence

- Description of the circuit in Figure 7.2, as described in the COMPACT manual.

- Creation of the computer connection, e.g. to the Cybernet computer network of Control Data, input of the circuit description created and checking its correctness.

- Call up and start COMPACT program, with output of the results.

- Disconnection of the computer link and, in this example, calculation of the costs.

(b) Circuit description

The preparation of a circuit description for COMPACT is described in Section 7.5.2 on the basis of numerous examples. A part-quadripole (e.g. SLC = series LC circuit or CAP = capacitance) is continuously identified by AA, BB etc, and its position in

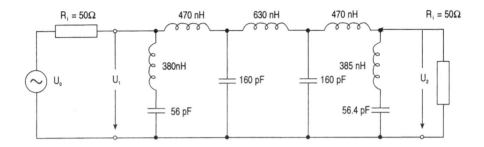

Figure 7.2 *Circuit for example*

the circuit is given (PA or SE = parallel or serial). This is followed by the input of the values (L in nH, C in pF, f in MHz). If all the part-quadripoles are given in the correct sequence, they are cascaded with

      CAX AA GG (connect all from AA to GG)

i.e. COMPACT calculates the ladder matrix of the resultant overall quadripole from the ladder matrices of the part quadripoles. With the command

      PRI AA VG 50 50 3 (VG = voltage gain; R_{in} = 50 ohm, R_{out} = 50 ohm)

COMPACT can be arranged to calculate the frequency-dependent attenuation function $U2/U1$, with its magnitude and phase being expressed in dB and degrees, and the group delay time being expressed in ns. The circuit description is terminated with END, and the input of the desired frequencies then follows. The frequency input is likewise terminated with END. A maximum of three frequency bands is possible. In this case,

15 30 2.5	(15–30 MHz in increments of 2.5 MHz)
31 35 1	(31–35 MHz in increments of 1 MHz)
37.5 40 45 50 60	(several individual frequencies)

The maximum number of frequencies is 30.

(c) Start

A session with COMPACT would proceed as follows:

| NEW, tpfilt | ◄────────── Open new file for the circuit description

READY.

| auto | ◄────────── Start input

```
00100 slc aa pa 380 56 197 25
00110 ind bb se 470 203 25
00120 cap cc pa 160
00130 ind dd se 630 224 25
00140 cap ee pa 160
00150 ind ff se 470 211 25
00160 slc gg pa 385 56.4 193 25
00170 caxaag
00180 pri aa vg 50 50 3
00190 end
00200 15 30 2.5
00210 31 35 1
00220 37.5 40 45 50 60
00230 end
```

Input prepared circuit description
line by line

00240 *DEL* ◄────────── After the last END, press Break or Interrupt key
(terminates input)

| 00170 cax aa gg |

00180 *DEL*   Line 00170 is input again because caxaagg was
keyed in without spaces.

READY

| save |

READY   Store the circuit file

save

| get, rfopt/un=vsslib | ◄─── Call up the RFOPT program

READY

| -rfopt | ◄────────── Start the RFOPT program

Name of file

FILE NAME, 'TRXINFO', 'TRLINES', 'LANGE', NEWS OR QUIT'?
tpfilt   AN(1), SENS(2), OPT(3), SV(4), MAP(5), VAR(6),
MC(7), 13(13), RND(44)?   1   EDIT YOUR INPUT FILE (Y/N)?

n

No editing   1 for analysis

VOLTAGE GAIN FROM 50.0 OHM SOURCE INTO 50.0 OHM LOAD

FREQ.	GAIN(DB)	PHASE(DEGREES)	DELAY
15.000	− 1.13	−140.44	28.88507
17.500	− .38	−168.99	34.90311
20.000	− .29	155.90	43.45318
22.500	− .18	112.86	51.01946
25.000	− 1.48	68.05	46.42381
27.500	− 2.80	29.46	42.85.090
30.000	− 1.82	− 87.61	327.89115
31.000	−10.29	−144.58	67.44868
32.000	−20.36	−160.12	29.31737
33.000	−30.71	−167.78	13.75802
34.000	−51.50	−154.95	−291.93298
35.000	−40.88	− 7.45	− 1.23332
37.500	−36.12	− 14.51	8.82463
40.000	−37.14	− 21.72	7.23029
45.000	−41.31	− 32.47	4.94456
50.000	−45.74	− 40.06	3.61687
60.0000	−53.79	− 50.21	2.20994

(RFOPT can be left by QUIT)

FILE NAME, 'TRXINFO', 'TRLINES', 'LANGE', 'NEWS' OR 'QUIT'?   quit

READY

bye

GI03012 LOG OFF 11.02.21.
SBU 4.708 CT 7.00   TIO = 2600.

7.6 SUPERCOMPACT program

7.6.1 Overview

SUPERCOMPACT is a CAE program which supports the development of high-frequency and microwave circuits in generating circuit design, accelerating development, and opening up new design possibilities. SUPERCOMPACT was completely redeveloped by Compact Engineering, drawing on the experience gained since 1973 with about 400 electronics companies throughout the world. The total software package has been in practical use since 1981 in a large number of companies.

SUPERCOMPACT consists of several high-performance program sections, and is very user-friendly. Typical applications are

- analysis, synthesis, and optimisation of linear circuits in the frequency range from the LF to the microwave sector

- concentrated and locally distributed elements (e.g. microstrip lines and long couplers)

- sensitivity analysis and Monte-Carlo tolerance analysis

- filters from coupled striplines

- passive and active lowpass, highpass and bandpass filters

- low-noise wideband amplifiers

- antenna matching circuits

- power combiners

- transistor models

- oscillator draft designs.

7.6.2 Improvements

Compared with COMPACT, SUPERCOMPACT provides a number of improvements:

- analysis and optimisation of three-port and four-port networks

- topology generation

- synthesis

- yield analysis

- editor substantially extended

- improved microstrip model

- additional optimisation procedures (random search)

- databank for dielectrics and part circuits

– screen graphics (Smith chart, interactive)

– mapping

– connection to SUPERFILSYN

– histograms

It provides the following *new elements:*

– spiral inductor

– interdigital capacitor

– distributed resistor

– bond wire element

And it provides the following *new properties:*

– algebraic links between variables (+, -, x,/,)

– BLK block may now contain a maximum of 50 nodes

– analysis section now with doubled precision

– error corrections

7.6.3 Program structure

The overall structure of SUPERCOMPACT is shown in Figure 7.3.

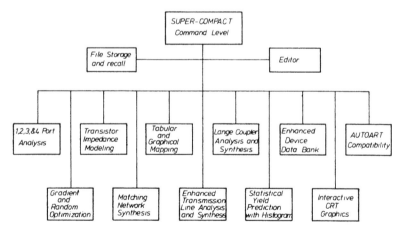

Figure 7.3   *Structure of SUPERCOMPACT*

7.6.4 Preparing circuit files

For SUPERCOMPACT the data takes the form of a sequence of input blocks as follows:

BLK
 .
 .   Circuit section as BLOCK
END

LAD
 .
 .   LADDER section, individual elements
END

NOI
 .
 .   Noise parameters
END

FREQ
 .
 .   Frequencies
END

OUT
 .
 .   Output data
END

OPT
 .
 .   Optimisation parameters
END

STAT
 .
 .   Statistics parameters
END

DATA
 .
 .   Data blocks required (e.g. S-parameters)
END

7.6.5 Elements for extraction

Passive two-port network (two-terminal lumped element) definitions and syntax are given in Figure 7.4.

Element	Syntax	Example
Resistance 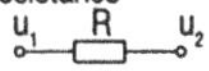 u_1 R u_2	RES n1 [n2] R = x	RES 3 R = 4.7 KOH
Inductance  u_1 L u_2	IND n1 [n2] L = x [Q = y F = z]	IND 3 8 L = 3.8 UH Q = 160 F = 7.9 MHz coil quality Q = 160 measured at 7.9 MHz
Capacitance 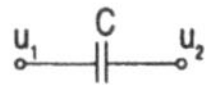 u_1 C u_2	CAP n1 [n2] C = x [Q = y F = z]	CAP 1 11 C 180PF Q = 1200 F 25 MHz capacitor quality Q = 1200 measured at F = 25 MHz
RL series connection  u_1 R L u_2	SRL n1 [n2] R = x, L = y	SRL 2 8 R = 10 L = 100 MHz
RL parallel connection 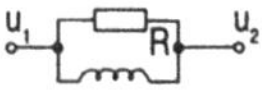 u_1 R u_2	PRL n1 [n2] R = x, L = y	PRL 2 8 R = 10 L = 100 MHz
RC series connection 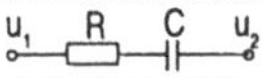 u_1 R C u_2	SRC n1 [n2] R = x, L = y	SRC 1 R = 10 C = 100 pF
RC parallel connection 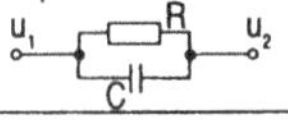 u_1 R u_2	PRC n1 [n2] R = x, L = y	PRC 5 15 R = 10 C = 100 pF

Figure 7.4 *Two-port element definitions for SUPERCOMPACT*

7.6.6 Synthesis of matching networks (PMS, SYN): overview

The PMS (port models and synthesis) program section is intended for the drafting of input and output matching networks for active components. With the PMS process, precise models for the active elements must first be determined by optimisation. Concentrated or line elements can then be synthesised with the aid of the SUPERCOMPACT synthesis program.

Before the PMS command can be used, the circuit file must be prepared in accordance with the rules described in the manual. The details of the theory on which this is based were developed by Medley and Allen, from whom the PMS process is derived, and who tested it out under practical conditions and proved its practical usability. It is conceivable that, in a future version of SUPERCOMPACT, the PMS procedure will be expanded to more than one stage.

7.6.7 Examples of circuit optimisation

Plot output

The following input file shows how circuit performance can be plotted. The output for this circuit is contained on the following pages.

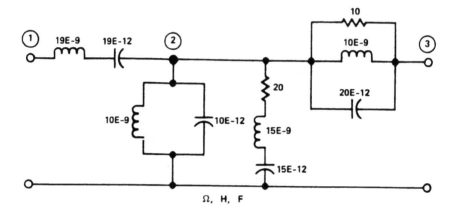

Figure 7.5   *Circuit example*

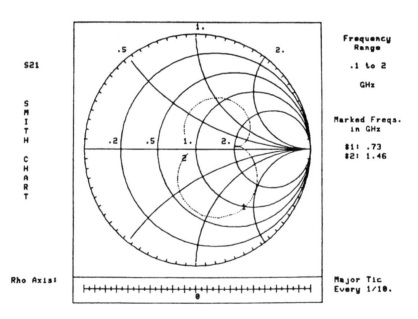

Figure 7.6   *Plot of* S₂₁ *for circuit example*

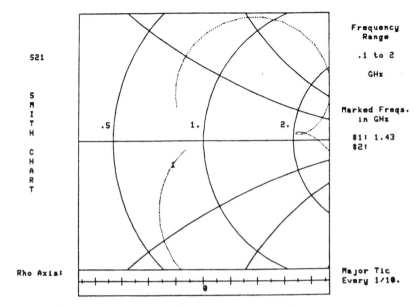

Figure 7.7   *Expanded plot of* S21 *for circuit example*

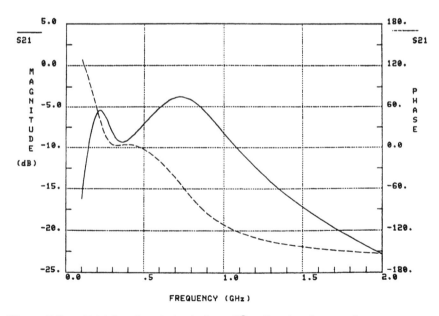

Figure 7.8   *Initial and optimised plots of* S21 *for circuit example*

Optimisation of input network of a power amplifier

This example (ex4) illustrates the use of optimisation to match a low impedance, the input impedance of a transistor, for example, to a 50 Ω source. Element IMP is used to represent the impedance to be matched. In the circuit description the IMP element is assigned label Q1. Note that IMP is both the name of the element and also a keyword in the Data section. The keyword IMP specifies that data for element IMP are impedances in real–imaginary (resistance–reactance) form. The single line in the Freq section, namely,

STEP 1GHz, 2GHz 200 MHz

specifies the set of frequencies

1.0, 1.2, 1.4, 1.6, 1.8, 2.0 GHz

In the Data section there are six lines of data. Because frequencies are not included in the Data section, the first data pair (1.9, 2.0) is assigned to the first frequency, 1.0 GHz, and likewise for the remaining five lines of data.

The circuit is defined as a one-port and is given the name A. The Opt section contains the specification.

A MS11 W 10

specifies that the magnitude of input reflection coefficient MS11 should be zero. By default, the reference is 50 Ω. A weight of 10 is specified but plays no essential role in the optimisation, since only one quantity is being optimised.

```
SUPER COMPACT  Version 1.5+20  12/01/81    15-DEC-81  10:10:20

CMD > ex4
CMD > LIS
* [COMPACT.TEST]EX4.DAT
* EXAMPLE #4, pages 34 and 35 of COMPACT User Manual, Version 5.1
*
LAD
 TRL 1 2 Z0=?35?   PLEN=?1.36IN? KEFF=2.1
 TRL 2 3 Z0 ?23?   PLEN ?1.26IN? KEFF 2.25
 CAP 3 0 C  ?20PF?
 TRL 3 4 Z0 ?12.5? PLEN ?0.16IN? KEFF 2.4
 CAP 4   C  ?20PF?
 IMP 4   Q1
A:1POR 1
END
FREQ
 STEP 1GHZ 2GHZ 200MHZ
END
OUT
 PRI A S
END
OPT
 A MS11 W 10
 TERM .2
END
DATA
Q1:IMP
 1.9 2.0
 2.0 2.1
 2.0 2.0
 1.9 1.8
 1.8 1.5
 1.7 1.0
END

CMD > ANA

CIRCUIT:  A
ONE-PORT, ZS =   50. +J   0.
 FREQ         RHO          SWR     RET L/G     OHMS         MILLIMHOS
 GHz      MAG     ANG               dB       R       X       G       B
1.00000  0.419  -53.2     2.44:1   -7.55   61.2   -49.8    9.8     8.0
1.20000  0.460  -98.8     2.70:1   -6.75   29.2   -33.6   14.7    17.0
1.40000  0.558 -140.3     3.52:1   -5.07   15.9   -16.4   30.4    31.5
1.60000  0.598 -179.3     3.97:1   -4.47   12.6    -0.3   79.4     1.9
1.80000  0.568  139.2     3.63:1   -4.92   15.5    17.0   29.3   -32.1
2.00000  0.450   84.8     2.64:1   -6.93   35.5    40.0   12.4   -14.0

PLOT, PRINT OR QUIT? (PL/PR/<Q>): Q
CMD > OPT
GRADIENT, RANDOM OR QUIT? (G/R/Q): G
<OPTIM> TIMER STARTS


OPTIMIZATION BEGINS WITH FOLLOWING VARIABLES AND GRADIENTS

     VARIABLES              GRADIENTS
( 1):   35.000          ( 1): -4.3821
( 2): 0.34544E-01       ( 2):-0.13590
( 3):   23.000          ( 3):  6.6495
( 4): 0.32004E-01       ( 4):-0.96798
( 5): 0.20000E-10       ( 5):  2.7490
( 6):   12.500          ( 6): -2.0289
( 7): 0.40640E-02       ( 7): -1.6308
( 8): 0.20000E-10       ( 8):  1.6761
ERR. F.=      2.634
      ----****----
```

MATCH TO 50Ω →

① 35Ω ② 23Ω ③ 12.5Ω ④
1360 mil   1260 mil   160 mil
20pf   20pf   IMP

```
NUMBER OF ITERATIONS? (X/<0>): 10
PRINT BEST VALUES EACH ITERATION? (Y/<N>): N
  ERR. F.=       0.362
  ( 1):   37.238              ( 1): 0.48205
  ( 2): 0.34105E-01           ( 2): 0.23581E-01
  ( 3):   15.619              ( 3):-0.36441          Note: The left-hand column gives element
  ( 4): 0.36767E-01           ( 4): 0.40088                values in [Ohms]. [Meters] and
  ( 5): 0.16345E-10           ( 5):-0.42021E-01             [Farads]
  ( 6):   15.433              ( 6):-0.67614E-02
  ( 7): 0.48463E-02           ( 7):-0.14342E-01
  ( 8): 0.17063E-10           ( 8):-0.75828E-01
  ERR. F.=       0.028
      ----****----
FUNCTION TERMINATION WITH ABOVE VALUES. FINAL ANALYSIS FOLLOWS

   TIMES IN SECONDS        PAGE     DIRECT    BUFFERED
    CPU     ELAPSED        FAULTS     I/O       I/O
    4.78     31.98           57        1        32

<OPTIM> CPU TIME =    4.78 SECS.
V = 2.7576E-02     72 FUNCTION EVALUATIONS       8 CALLS TO <COSTT>


CIRCUIT:  A
ONE-PORT, ZS =   50. +J    0.
  FREQ         RHO            SWR     RET L/G    OHMS          MILLIMHOS
   GHz     MAG     ANG                 dB      R      X       G       B
  1.00000  0.018 -118.3     1.04:1   -34.95   49.1   -1.5    20.3    0.6
  1.20000  0.079   74.8     1.17:1   -22.08   51.5    7.9    19.0   -2.9
  1.40000  0.046    2.2     1.10:1   -26.83   54.8    0.2    18.3   -0.1
  1.60000  0.033  -91.2     1.07:1   -29.68   49.8   -3.3    20.0    1.3
  1.80000  0.004  147.8     1.01:1   -49.05   49.7    0.2    20.1   -0.1
  2.00000  0.083  -18.1     1.18:1   -21.63   58.5   -3.0    17.1    0.9

PLOT, PRINT OR QUIT? (PL/PR/<Q>): Q


!!ON RETURN TO 'CMD> ' TYPE 'SAVE' AND AN OPTIONAL FILE NAME TO SAVE OPTIMIZED
  RESULTS -- EXISTING FILE IS OVERWRITTEN, IF FILE NAME IS NOT SUPPLIED!!

CMD > SAVE EX4OPT


CMD > QUI

*** END SUPER COMPACT ***

FORTRAN STOP
$ TY EX4OPT.DAT
* [COMPACT.TEST]EX4.DAT
* EXAMPLE #4, pages 34 and 35 of COMPACT User Manual, Version 5.1
*
LAD
 TRL 1 2 Z0=?37.238?    PLEN=?1.3427IN? KEFF=2.1
 TRL 2 3 Z0 ?15.619?    PLEN ?1.4475IN? KEFF 2.25
 CAP 3 0 C  ?16.345PF?
 TRL 3 4 Z0 ?15.433? PLEN ?.1908IN? KEFF 2.4
 CAP 4    C  ?17.063PF?
 IMP 4    Q1
A:1POR 1
END
FREQ .
 STEP 1GHZ 2GHZ 200MHZ
END
OUT
 PRI A S
END
```

```
OPT
 A MS11 W 10
 TERM .2
END
DATA
Q1:IMP
 1.9 2.0
 2.0 2.1
 2.0 2.0
 1.9 1.8
 1.8 1.5
 1.7 1.0
END
$
```

Optimisation of 10–1000 MHz wideband amplifier

EX8 shows the optimisation of a single stage amplifier. The OPT section specifies perfect input and output match (MS11=0, MS22=0) and flat gain of 10 dB over the frequency band from 10 MHz to 1 GHz.

```
SUPER COMPACT  Version 1.5+20  12/01/81     15-DEC-81  14:58:17

CMD > EX8
CMD > LIS
* [COMPACT.TEST]EX8.DAT
* EXAMPLE #8, page 41 of COMPACT User Manual, Version 5.1
* All nodal
*
BLK
 CAP 1 0 C=?4PF?
 IND 1 2 L ?15NH?
 RES 2   R 550
 TWO 2 3 4 Q1
 PRC 4   R ?5? C ?5PF?
 SRL 2 3 R ?400? L ?40NH?
 RES 3   R 300
A:2POR 1 3
END
FREQ
 10MHZ STEP 250MHZ 1000MHZ 250MHZ
END
OUT
 PRI A S
END
OPT
 A MS11 MS22 MS21 10DB W 2
 TERM 0.1
END
DATA
Q1:S
 .42  -40 35  170 .005 80 .9  -10
 .50  -75 21  145 .02  65 .7  -18
 .58 -120 9.4 121 .04  40 .52 -28
 .48 -153 6.2 109 .04  45 .49 -29
 .58 -144 4.5 102 .05  49 .46 -30
END
CMD > ANA
```

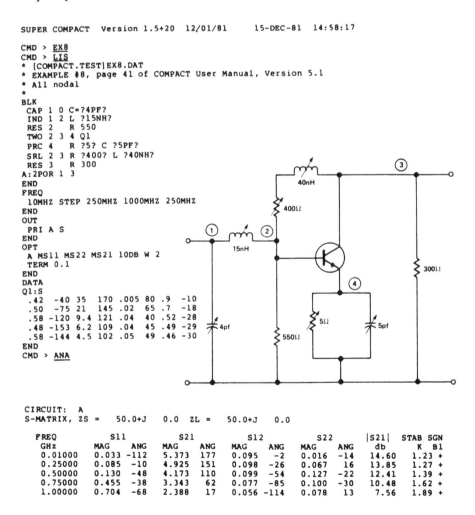

```
CIRCUIT:  A
S-MATRIX, ZS =   50.0+J   0.0  ZL =   50.0+J   0.0
```

FREQ	S11		S21		S12		S22		|S21|	STAB SGN	
GHz	MAG	ANG	MAG	ANG	MAG	ANG	MAG	ANG	db	K	B1
0.01000	0.033	-112	5.373	177	0.095	-2	0.016	-14	14.60	1.23	+
0.25000	0.085	-10	4.925	151	0.098	-26	0.067	16	13.85	1.27	+
0.50000	0.130	-48	4.173	110	0.099	-54	0.127	-22	12.41	1.39	+
0.75000	0.455	-38	3.343	62	0.077	-85	0.100	-30	10.48	1.62	+
1.00000	0.704	-68	2.388	17	0.056	-114	0.078	13	7.56	1.89	+

```
PLOT, PRINT OR QUIT? (PL/PR/<Q>): Q
CMD > OPT
GRADIENT, RANDOM OR QUIT? (G/R/Q): G
<OPTIM> TIMER STARTS

OPTIMIZATION BEGINS WITH FOLLOWING VARIABLES AND GRADIENTS

      VARIABLES                  GRADIENTS
   ( 1): 0.40000E-11          ( 1):  6.0654
   ( 2): 0.15000E-07          ( 2):  14.820
   ( 3):  5.0000             ( 3): -31.700
   ( 4): 0.50000E-11          ( 4):-0.20981
   ( 5):  400.00             ( 5):  31.700
   ( 6): 0.40000E-07          ( 6): -1.8120
  ERR. F.=     19.354
       ----****----
NUMBER OF ITERATIONS? (X/<0>): 10 N
  ERR. F.=      0.588
  ERR. F.=      0.170
   ( 1): 0.33913E-11          ( 1): 0.91344E-01
   ( 2): 0.10744E-07          ( 2):-0.13858E-01
   ( 3):  8.0319             ( 3): 0.18224
   ( 4): 0.52287E-11          ( 4):-0.15944E-01
   ( 5):  230.83             ( 5): 0.31665E-01
   ( 6): 0.47295E-07          ( 6):-0.79200E-01
  ERR. F.=      0.089
       ----****----

FUNCTION TERMINATION WITH ABOVE VALUES. FINAL ANALYSIS FOLLOWS

  TIMES IN SECONDS         PAGE     DIRECT   BUFFERED
   CPU     ELAPSED        FAULTS     I/O       I/O
   5.90    32.24            91        1        27

<OPTIM> CPU TIME =    5.90 SECS.
V = 8.8685E-02     77 FUNCTION EVALUATIONS     11 CALLS TO <COSTT>

CIRCUIT:  A
S-MATRIX, ZS =   50.0+J   0.0  ZL =   50.0+J   0.0
```

FREQ	S11		S21		S12		S22		|S21|	STAB SGN	
GHz	MAG	ANG	MAG	ANG	MAG	ANG	MAG	ANG	db	K	B1
0.01000	0.066	-167	3.206	178	0.146	-2	0.064	178	10.12	1.30	+
0.25000	0.086	38	3.150	166	0.146	-26	0.093	95	9.97	1.29	+
0.50000	0.205	-47	3.097	138	0.135	-54	0.166	39	9.82	1.32	+
0.75000	0.263	-58	3.168	103	0.103	-78	0.228	14	10.02	1.49	+
1.00000	0.269	-75	3.179	67	0.078	-98	0.233	-3	10.05	1.84	+

```
PLOT, PRINT OR QUIT? (PL/PR/<Q>): Q

!!ON RETURN TO 'CMD> ' TYPE 'SAVE' AND AN OPTIONAL FILE NAME TO SAVE OPTIMIZED
   RESULTS -- EXISTING FILE IS OVERWRITTEN, IF FILE NAME IS NOT SUPPLIED!!

CMD > END

*** END SUPER COMPACT ***
```

Optimisation of microwave filter

EX3 uses 16 quarter-wave resonators cascaded to from a bandpass filter that has mirror symmetry about the centre. One half is formed in the LAD section and this half is named B; it is turned end-for-end and cascaded with itself in the BLK section to form the complete filter.

This example includes an element (resonator A) that is used twice within the LAD section of the circuit description. To use a circuit element more than one time within the same BLK, LAD or NOI section it is only necessary to assign a user defined label, a one-to-eight character variable followed immediately by a colon ":", to the element the first time that element appears. Subsequently, this labelled circuit element can be re-used by including the label with appropriate node numbers. Such usage is valid only within the same BLK, LAD or NOI section. Invalid results can be obtained if an element defined within a LAD section is used within a BLK section.

A very important point in this example is that only in the BLK section can the halfsection B be turned end-for-end. In a LAD or NOI section the order of node numbers assigned to a previously defined circuit conveys no information about direction. In the example:

```
LAD             LAD
  B 1 2           B 1 2
  B 3 2           B 2 3
  C:2POR 1 3      D:2POR 1 3
End             End
```

the two ports C and D so defined are identical.

EX3B.DAT is used for optimisation.

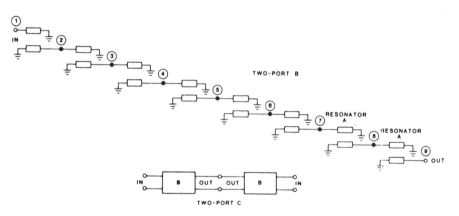

```
SUPER COMPACT   Version 1.5+20   12/01/81      15-DEC-81  10:24:19

CMD > EX3
CMD > LIS
*  [COMPACT.TEST]EX3.DAT
*
F1:14.34GHZ
LAD
  QLB  1  2  F=F1  Z0=52.5    Z1=44.4    Z2=132.4
  QLB  2  3  F  F1  Z0  62.5   Z1  132.4   Z2  131.2
  QLB  3  4  F  F1  Z0  68.4   Z1  131.2   Z2  126
  QLB  4  5  F  F1  Z0  78.0   Z1  126     Z2  125
  QLB  5  6  F  F1  Z0  78.0   Z1  126     Z2  125
  QLB  6  7  F  F1  Z0  58.0   Z1  126     Z2  125
A:QLB  7  8  F  F1  Z0  58.0   Z1  126     Z2  125
  A    8  9
B:2POR  1  9
END
```

```
BLK
 B 1 2
 B 3 2
C:2POR 1 3
END
FREQ
 STEP  6.5GHZ     7.5GHZ   500MHZ
 STEP  8.675GHZ 18.675GHZ  2GHZ
 STEP 21GHZ      22GHZ     500MHZ
END
OUT
 PRI C S
END
CMD > ANA


CIRCUIT:  C
S-MATRIX, ZS =    50.0+J   0.0  ZL =    50.0+J    0.0
```

FREQ	S11		S21		S12		S22		|S21|	STAB SGN	
GHz	MAG	ANG	MAG	ANG	MAG	ANG	MAG	ANG	db	K	B1
6.50000	1.000	110	0.000	20	0.000	20	1.000	110	-73.51	2.67	-
7.00000	1.000	97	0.001	7	0.001	7	1.000	97	-57.93	1.04	-
7.50000	1.000	77	0.018	-13	0.018	-13	1.000	77	-34.91	1.00	-
8.67500	0.211	34	0.977	-56	0.977	-56	0.211	34	-0.20	1.00	+
10.67500	0.064	-30	0.998	-120	0.998	-120	0.064	-30	-0.02	1.00	+
12.67500	0.316	171	0.949	-99	0.949	-99	0.316	171	-0.46	1.00	-
14.67500	0.021	-143	1.000	-53	1.000	-53	0.021	-143	0.00	1.00	-
16.67500	0.172	81	0.985	-9	0.985	-9	0.172	81	-0.13	1.00	-
18.67500	0.664	-91	0.748	-1	0.748	-1	0.664	-91	-2.52	1.00	-
21.00000	0.994	-62	0.108	28	0.108	28	0.994	-62	-19.31	1.00	-
21.50000	1.000	-91	0.003	-1	0.003	-1	1.000	-91	-50.99	1.00	-
22.00000	1.000	-106	0.000	-16	0.000	-16	1.000	-106	-68.38	0.00	-

```
PLOT, PRINT OR QUIT? (PL/PR/<Q>): Q
CMD > END

*** END SUPER COMPACT ***


 SUPER COMPACT  Version 1.5+20  12/01/81    15-DEC-81  10:29:12

CMD > EX3B
CMD > LIS
* [COMPACT.TEST]EX3B.DAT
*
F1:14.34GHZ
LAD
  QLB 1 2 F=F1 Z0=?52.5?  Z1=?44.4?  Z2=132.4
  QLB 2 3 F F1 Z0 ?62.5?  Z1 ?132.4?· Z2 131.2
  QLB 3 4 F F1 Z0 ?68.4?  Z1 ?131.2? Z2 126
  QLB 4 5 F F1 Z0 ?78.0?  Z1·?126?   Z2 125
  QLB 5 6 F F1 Z0 ?78.0?  Z1 ?126?   Z2 125
  QLB 6 7 F F1 Z0 ?80.58? Z1 ?126?   Z2 125
A:QLB 7 8 F F1 Z0 ?72.25? Z1 ?107?   Z2 125
  A   8 9
B:2POR 1 9
END
BLK
 B 1 2
 B 3 2
C:2POR 1 3
END
FREQ
 STEP 6.5GHZ     7.5GHZ   500MHZ
 STEP 8.675GHZ 18.675GHZ  2GHZ
 STEP 21GHZ     22GHZ     500MHZ
END
OPT
 C F 6.5GHZ     7.5GHZ    MS21 -40DB LT
   F 8.675GHZ 18.675GHZ  MS21   0DB
   F 21GHZ     22GHZ     MS21 -40DB LT
END
```

```
OUT
 PRI C S
END
CMD > OPT
GRADIENT, RANDOM OR QUIT? (G/R/Q): G
<OPTIM> TIMER STARTS


OPTIMIZATION BEGINS WITH FOLLOWING VARIABLES AND GRADIENTS

       VARIABLES                   GRADIENTS
    ( 1):    52.500           ( 1):-0.46534E-01
    ( 2):    44.400           ( 2):-0.16210E-01
    ( 3):    62.500           ( 3):-0.29881E-01
    ( 4):   132.40            ( 4): 0.62725E-02
    ( 5):    68.400           ( 5): 0.61961E-01
    ( 6):   131.20            ( 6):-0.41397E-02
    ( 7):    78.000           ( 7): 0.57183E-02
    ( 8):   126.00            ( 8):-0.72923E-02
    ( 9):    78.000           ( 9):-0.95186E-01
    (10):   126.00            (10): 0.82050E-02
    (11):    80.580           (11): 0.17213
    (12):   126.00            (12):-0.86846E-02
    (13):    72.250           (13):-0.11480
    (14):   107.00            (14): 0.16186E-01
    ERR. F.=        0.005
         ----****----
NUMBER OF ITERATIONS? (X/<0>): 10 N
    ERR. F.=        0.001
    ERR. F.=        0.000
    ERR. F.=        0.000
    ( 1):    53.515           ( 1):-0.22812E-02
    ( 2):    44.915           ( 2):-0.80021E-03
    ( 3):    61.373           ( 3):-0.56578E-03
    ( 4):   132.45            ( 4): 0.28129E-03
    ( 5):    69.700           ( 5): 0.59583E-03
    ( 6):   131.03            ( 6): 0.10834E-03
    ( 7):    79.433           ( 7):-0.96716E-03
    ( 8):   125.27            ( 8): 0.19325E-03
    ( 9):    77.787           ( 9):-0.36525E-02
    (10):   125.80            (10): 0.67877E-03
    (11):    75.261           (11): 0.48322E-02
    (12):   127.12            (12):-0.21755E-04
    (13):    74.552           (13):-0.18413E-02
    (14):   106.60            (14): 0.36343E-04
    ERR. F.=        0.000
         ----****----
GRADIENT TERMINATION WITH ABOVE VALUES. FINAL ANALYSIS FOLLOWS



   TIMES IN SECONDS        PAGE     DIRECT   BUFFERED
    CPU    ELAPSED        FAULTS     I/O      I/O
   41.55   120.42           381       1        44

<OPTIM> CPU TIME =   41.55 SECS.
V = 5.5765E-05    195 FUNCTION EVALUATIONS    13 CALLS TO <COSTT>

CIRCUIT:  C
S-MATRIX, ZS =   50.0+J   0.0  ZL =   50.0+J   0.0

  FREQ        S11          S21          S12          S22       |S21|  STAB SGN
  GHz     MAG   ANG    MAG   ANG    MAG   ANG    MAG   ANG     db    K   B1
  6.50000 1.000 110   0.000  20   0.000  20   1.000 110   -84.95 -18.61 +
  7.00000 1.000  97   0.000   7   0.000   7   1.000  97   -71.58   1.71 -
  7.50000 1.000  78   0.002 -12   0.002 -12   1.000  78   -54.72   0.97 +
  8.67500 0.027 120   1.000  30   1.000  30   0.027 120     0.00   1.00 +
 10.67500 0.018 175   1.000 -95   1.000 -95   0.018 175     0.00   1.00 -
 12.67500 0.025   3   1.000 -87   1.000 -87   0.025   3     0.00   1.00 +
 14.67500 0.052 -144  0.999 -54   0.999 -54   0.052 -144   -0.01   1.00 -
 16.67500 0.050 -117  0.999 -27   0.999 -27   0.050 -117   -0.01   1.00 -
 18.67500 0.366 -126  0.931 -36   0.931 -36   0.366 -126   -0.62   1.00 +
 21.00000 1.000 -69   0.004  21   0.004  21   1.000 -69   -47.31   1.00 -
 21.50000 1.000 -91   0.001  -1   0.001  -1   1.000 -91   -66.01   0.48 -
 22.00000 1.000 -105  0.000 -15   0.000 -15   1.000 -105  -80.44 -13.20 -

PLOT, PRINT OR QUIT? (PL/PR/<Q>): Q
GRADIENT, RANDOM OR QUIT? (G/R/Q): Q


!!ON RETURN TO 'CMD> ' TYPE 'SAVE' AND AN OPTIONAL FILE NAME TO SAVE OPTIMIZED
  RESULTS -- EXISTING FILE IS OVERWRITTEN, IF FILE NAME IS NOT SUPPLIED!!

CMD > END

*** END SUPER COMPACT ***
```

7.7 SUPERFILSYN program

7.7.1 Overview

SUPERFILSYN is a program for active, passive, digital, and microwave filters. The user provides the filter specifications at the terminal, and S/FILSYN determines from this the transfer function which will fulfil the requirements. This function can then be implemented in the next stage as a passive LC, active RC, digital (IIR or FIR) or microwave filter. For passive LC and microwave filters, an automatic determination of the filter structure is possible, with the output of the element values. In the case of active filters, the part blocks can be completely dimensioned.

S/FILSYN is suitable for the following types of filter:
– lowpass
– highpass
– linear phase lowpass
– bandpass
– bandstop

Attenuation behaviour, as a maximum, flat or uniform (i.e. Chebyshev) can be required. An existing input of a transfer function is possible.

Advantages with S/FILSYN

– The program is set up in such a way that it can be used equally well by nonspecialists and experts. The program application is described completely and the user dialogue is designed in such a way that only those questions are asked which are necessary for the particular filter type concerned.

– Standard filters can be rapidly and reliably dimensioned by any user. Conversion calculations which are prone to error, such as lowpass/bandpass transformation or normalisation and denormalisation are no longer required.

– Expenditure estimates and alternative solution possibilities can be easily worked out.

– A complete analysis of the filter (attenuation and phase, group delay, impedance values) is possible, and, together with the protocol of the program flow, provides comprehensive documentation for the filter which has been drafted. The graphic output allows for an at-a-glance appraisal of the filter properties.

7.7.2 Program structure

Figure 7.9 shows the program structure of S/FILSYN. The program flow can start at the point marked E. Within the segments, control is effected by the dialogue between program and user. The PLACER segment is required if the transfer function has to be determined iteratively by the placement of the poles; in other words, for example, with the following requirements:

– section-by-section constant attenuation in the stopband area
– parametric bandpasses with various stopband points
– microwave filters with λ/4 line sections and end stopband points.

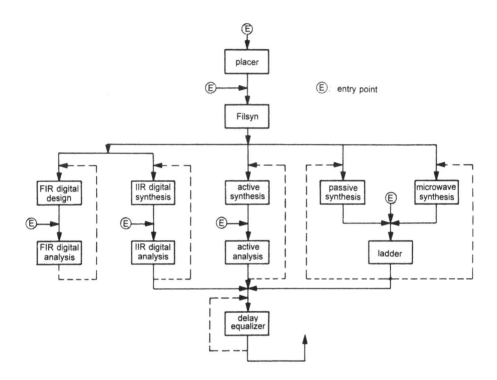

Figure 7.9   *General flowchart for S/FILSYN*

If filters are to be designed for which a closed solution exists, a start can be made immediately at the FILSYN segment with the input of the specifications. After FILSYN has run the realisation segment required can be executed (e.g. active synthesis). If a synthesised circuit already exists, it can be examined in the individual analysis segment concerned, and, if necessary, modified interactively.

7.7.3 Typical areas of application

7.7.3.1 Passive LC and microwave filters

Possible structures include general ladder circuits, which may contain bridged-T elements, and cross members for symmetrical LC filters. The synthesis stage for determining the network structure and the element values can be carried out either automatically by the program or initiated by the user. In the latter type of application, a choice can be made between three levels, based on experience. The filter structure which is evaluated can be analysed in the frequency range with or without consideration of the losses. An extensive command set allows for changes in the filter structure, such as impedance transformations, delay time distortion, application of Kuroda identities for microwave filters, determination of the dual structure, conversions of ladder circuits into cross-elements, and frequency or impedance scaling.

S/FILSYN can also create an input file for SUPER-COMPACT. If this program is available, it is possible to carry out a statistical analysis or optimisation of the filter; this direct data exchange enables errors to be avoided and saves time.

7.7.3.2 Active RC filters

These can be implemented with S/FILSYN in multistage technology (i.e. as a cascade of stages of first and second order), or in leapfrog or follow-the-leader form. The program makes it possible to carry out a frequency dependent analysis of the individual structure, the part blocks of which can be realised with operational amplifiers, with the network being represented on-screen. In addition, delay time distortion and other analyses are possible.

7.7.3.3 Digital filters

Nonrecursive or FIR (finite impulse response) filters can be developed in linear-phase form with up to **ten** flow and block areas. The individual filter in each case can be analysed and processed with S/FILSYN, with, for example, clipping, factorising, and conversion into minimum phase functions all being possible. Recursive or IIR (infinite impulse response) filters can be created starting from the analogue transfer functions by means of the impulse-invariant, matched, or bilinear z-transformation. The filter which is developed can be represented in cascade and/or parallel form, analysed in time and/or frequency range and modified many times over. In this context, for example, overflow and rounding, scaling with several methods, pairing and ordering, permutating and equalising of the group delay are all possible.

7.7.3.4 Filter characteristics

Table 7.1 provides a series of current filter characteristics which can be produced with S/FILSYN.

7.7.3.5 Additional properties and application

All analysis and processing segments of S/FILSYN can be used as independent subprograms for already existing filter structures, i.e. not produced with S/FILSYN. The dialogue can be interrupted at any time during processing. The part result obtained is stored in a file by the program, and read in again when the dialogue is continued. The data input is simple and is carried out in free format. The interactive dialogue allows for flexible control with the next command in each case being input after evaluation of the intermediate results obtained. A HELP function is available for command explanation. S/FILSYN is written completely in FORTRAN and to a large extent is platform-independent.

Table 7.1   *Available filter characteristics*

Characteristics	Attenuation requirements DB	Group delay requirements SB	DB
Butterworth	Maximally flat	monotonic	–
Chebyshev	Constant ripple	monotonic	–
Chebyshev (inverted)	Maximally flat	constant minimum attenuation	–
Elliptical (Cauer)	Maximally flat	constant minimum attenuation	–
Bessel	–	random	maximally flat
Bessel (modified)	–	constant minimum attenuation	maximally flat
Ulbrich-	–	random	const. ripple
Piloty	–	constant minimum attenuation	const. ripple
General	maximum 1 x constant ripple against transfer function	random (PLACER) (PLACER) (PLACER)	– – –

7.7.4 Examples

Cauer highpass filter

A passive highpass filter is to be designed from lumped elements with a cut-off frequency of 7.5 MHz. It is intended that a reflectance factor of 20% (corresponding to 0.1773 dB) should be maintained in the passband. The guaranteed minimum attenuation in the stopband from 6 MHz should be 40 dB. The filter is intended to be connected at the input and output with 50 Ω. The solution with S/FILSYN follows.

```
         ***** S/FILSYN *****

RELEASE 1.0   VERSION 01    01/01/82

PLACER: P,FILSYN: F,LADDER: L,DIGITAL: D,ACTIVE: A OR END: E
> F
ENTER TITLE
  CAUER-HOCHPASS
FILTER KIND - LUMPED: O,DIGITAL: 1 OR MICROWAVE: 2
> 0
FILTER TYPE - LOWPASS: 1, HIGHPASS: 2, LIN.-PHASE LOWPASS: 3,
                                            BANDPASS: 4
  2
THE PASSBAND IN HZ
> 7.5 E6
PASSBAND - MAX.-FLAT: =, EQUAL-RIPPLE: 1, FUNCTIONAL INPUT: 2
> 1
WHAT IS THE BAND EDGE LOSS IN DB
> 0.1773
STOPBAND - MONOTONIC: 0, EQUAL-MINIMA: 1 OR SPECIFIED: 2
> 1
WISH TO SPECIFY MINIUM REQUIRED STOPBAND LOSS: Y/N
> Y
ENTER LOSS IN DB
> 40
ENTER EDGE FREQUENCY OF LOWER STOPBAND IN HZ
> 6E6
ENTER INPUT TERMINATION IN OHMS
> 50
ENTER OUTPUT TERMINATION (O. INDICATES OPEN OR SHORT)
> 50
ENTER VALUE OF AVERAGE Q. IF NO PREDISTORTION, ENTER O.
> 0

GENERAL FILTER SYNTHESIS PROGRAM
```

```
CAUER-HOCHPASS
  HIGH-PASS FILTER
    EQUAL RIPPLE PASS BAND
      BANDEDGE LOSS                            = 0.1773 DB.
      LOWER PASSBAND EDGE FREQUENCY            = 7.5000000D+06 HZ.
    EQUAL MINIMA STOP BAND WITH EDGE FREQUENCY= 6.0000000D+06 HZ.
      REQUIRED STOP BAND LOSS                  =   40.00 DB.
      MULTIPLICITY OF ZERO AT ZERO            = 2
      NUMBER OF FINITE TRANSMISSION ZERO PAIRS=2
      OVERALL FILTER DEGREE                   = 6
      TRANSMISSION ZEROS

              REAL PART          IMAGINARY PART

          0.0000000D+00        5.9956535D+06
          0.0000000D+00        4.7388373D+06
      TRANSMISSION ZEROS HAVE BEEN SHIFTED, NEW VALUES ARE


              REAL PART          IMAGINARY PART

          0.0000000D+00        5.8748307D+06
          0.0000000D+00        4.5829851D+06
      INPUT TERMINATION                        = 5.0000000D+01OHMS
      OUTPUT TERMINATION                       = 5.0000000D+01OHMS
      REQUESTED TERMINATION RATIO              = 1.0000000D+00
      NEAREST AVAILABLE TERMINATION RATIO      = 1.0000000D+00

WISH TO SEE TRANSFER FUNCTION: Y/N
> N
ENTER THE NUMBER OF INCREMENTS (UP TO 5).NO ANALYSIS, ENTER: 0
> 0
REALIZATION - ACTIVE: A, PASSIVE: P, DIGITAL: D, NO SYNTHESIS: E
 > P
EXISTING POLYNOMIALS ARE: ES OS ED
ENTER SUBTITLE. NO MORE SYNTHESES: END
 > CAUER-HOCHPASS PASSIVE REALISIERUNG
LATTICE: L, COMPUTER CONFIG.: C, INPUT SIDE: IN, OUTPUT SIDE: OUT
 > C
WISH TO SEE INTERMEDIATE RESULTS: Y/N
 > N
** EVEN NUMBERED BRANCHES ARE SERIES, ODD ONES SHUNT **

CAUER-HOCHPASS PASSIVE REALISIERUNG
```

```
1      ....R....      5.0000000D+01

       .        .

3      ....L....      9.3073446D-07

       .        .

4      .        C      3.1048917D-10

       .        .

5      ...L.C...      7.6444540D-07      RES.FREQUENCY

       .        .      1.5776000D-09    4.5829851D+06

       .        .

6      .        C      3.2250835D-10

       .        .                .

7      ...L.C...      1.3789492D-06      RES.FREQUENCY

       .        .      5.3223236D-10    5.8748307D+06

       .        .

8      .        C      6.9063111D-10

       .        .

9      ....R....      5.0000000D+01
```

COMMAND:

> FREQ

ENTER THE NUMBER OF FREQUENCY RANGES (NOT MORE THAN 5)

> 3

ENTER 3 FREQUENCY INCREMENTS

> 200E3   100E3   200E3

ENTER 4 CORNER FREQUENCIES

> 4E6   6E6   7.6E6   9E6

ENTER QL AND QC AT (UPPER) CUTOFF (LOSSLESS: ENTER ZEROS)

> 165 1250

CONSTAND Q'S: 1, OR CONSTANT DISSIPATION FACTORS: -1

> 1

TABULATE: Y/N

> Y

******* COMPUTED PERFORMANCE *******

FREQUENCY IN HZ	TRANSD. LOSS IN DB	PHASE IN DEG	DELAY IN SEC	OUTPUT IMPEDANCE REAL	IMAGINARY
4.0000E+06	45.7790	238.1141	4.2714E-08	2.1701E-01	-8.8208E+01
4.2000E+06	48.6692	240.9786	3.5039E-03	2.2325E-01	-8.1911E+01
4.4000E+06	54.4344	242.2294	2.2725E-04	2.3123E-01	-7.6021E+01
4.6000E+06	71.9248	113.7299	2.1206E-04	2.4120E-01	-7.0467E+01
4.8000E+06	52.2189	79.3949	4.7463E-09	2.5347E-01	-6.5189E+01
5.0000E+06	46.5809	81.7331	4.6633E-08	2.6836E-01	-6.0131E+01
5.2000E+06	43.6328	85.5164	5.6365E-08	2.8622E-01	-5.5240E+01
5.4000E+06	42.3997	89.6841	5.7829E-08	3.0756E-01	-5.0464E+01
5.6000E+06	43.3313	93.4064	3.8915E-08	3.3334E-01	-4.5744E+01
5.8000E+06	50.8224	88.6514	1.7201E-04	3.6576E-01	-4.1016E+01
6.0000E+06	43.1406	299.8226	1.6656E-04	4.1052E-01	-3.6194E+01
6.1000E+06	36.4501	299.6433	4.6221E-08	4.4154E-01	-3.3713E+01
6.2000E+06	31.6127	302.2354	8.9187E-08	4.8264E-01	-3.1161E+01
6.3000E+06	27.6235	305.9345	1.1327E-07	5.3974E-01	-2.8513E+01
6.4000E+06	24.1148	310.4033	1.3330E-07	6.2269E-01	-2.5739E+01
6.5000E+06	20.9097	315.5977	1.5398E-07	7.4803E-01	-2.2798E+01
6.6000E+06	17.9105	321.5922	1.7784E-07	9.4393E-01	-1.9637E+01
6.7000E+06	15.0620	328.5399	2.0685E-07	1.2591E+00	-1.6188E+01
6.8000E+06	12.3387	336.6615	2.4274E-07	1.7788E+00	-1.2361E+01
6.9000E+06	9.7452	346.2303	2.8662E-07	2.6551E+00	-8.0509E+00
7.0000E+06	7.3226	357.5166	3.3700E-07	4.1586E+00	-3.1488E+00
7.1000E+06	5.1552	10.6447	3.8709E-07	6.7618E+00	2.3796E+00
7.2000E+06	3.3579	25.3586	4.2329E-07	1.1219E+01	8.2824E+00
7.3000E+06	2.0256	40.8572	4.3089E-07	1.8434E+01	1.3553E+01
7.4000E+06	1.1671	56.0037	4.0694E-07	2.8543E+01	1.5956E+01
7.5000E+06	0.6933	69.8564	3.6320E-07	3.9187E+01	1.3074E+01
7.6000E+06	0.4733	82.0047	3.1491E-07	4.6154E+01	5.6309E+00
7.8000E+06	0.3802	101.5544	2.3584E-07	4.6165E+01	-8.0518E+00
8.0000E+06	0.4036	116.5363	1.8600E-07	4.0902E+01	-1.2300E+01
8.2000E+06	0.4101	128.7072	1.5532E-07	3.7493E+01	-1.1739E+01
8.4000E+06	0.3847	139.1007	1.3521E-07	3.6170E+01	-9.6577E+00
8.6000E+06	0.3396	148.2794	1.2081E-07	3.6308E+01	-7.2644E+00
8.8000E+06	0.2882	156.5538	1.0968E-07	3.7439E+01	-5.0124E+00
9.0000E+06	0.2397	164.1037	1.0048E-07	3.9256E+01	-3.1029E+00

The pass attenuation is not attained because of poor component quality.

Active bandpass

An active bandpass with a passband range from 885 to 1300 Hz (3 dB frequencies) is to be developed with S/FILSYN. The filter degree $n = 4$ is determined by the zeros, twofold in each case, at zero and infinite frequencies. The transfer function is to be implemented by two stages (each of order 2).

The solution with S/FILSYN follows.

```
                    *****S/FILSYN*****

      RELEASE 1.0  VERSION 01   01/01/82

      PLACER: P, FILSYN: F, LADDER: L, DIGITAL: D, ACTIVE: A OR END: E
      >F
      ENTER TITLE
        AKTIVER BANDPASS 885-1300 Hz
      FILTER KIND - LUMPED: O, DIGITAL: 1 OR MICROWAVE: 2
      >0
      FILTER TYPE - LOWPASS:  1, HIGHPASS: 2, LIN.-PHASE LOWPASS:
                              3, BANDPASS: 4
      >4
      LOWER EDGE OF THE PASSBAND IN HZ
      >885
      UPPER EDGE OF THE PASSBAND IN HZ
      >1300
      PASSBAND - MAX.-FLAT: 0, EQUAL-RIPPLE: 1, FUNCTIONAL INPUT: 2
      >1
      WHAT IS THE BAND EDGE LOSS IN DB
      >3
      BANDPASS - CONVENTIONAL: 1, PARAMETRIC: 2 OR MATCHING: 3
      >1
      STOPBAND - MONOTONIC: 0, EQUAL-MINIMA: 1 OR SPECIFIED: 2
      >2
      ENTER MULTIPLICITY OF TRANSMISSION ZERO AT ZERO
      >2
      ENTER MULTIPLICITY OF TRANSMISSION ZERO AT INFINITY
      >2
      ENTER NUMBER OF FINITE TRANSMISSION ZEROS
      >0

      ENTER INPUT TERMINATION IN OHMS
      >1
      ENTER OUTPUT TERMINATION (O. INDICATES OPEN OR SHORT)
      >0
      ENTER VALUE OF AVERAGE Q. IF NO PREDISTORTION, ENTER O.
      >0
```

```
GENERAL FILTER SYNTHESIS PROGRAM

AKTIVER BANDPASS 885-1300 Hz
  BAND-PASS FILTER
    EQUAL RIPPLE PASS BAND
      BANDEDGE LOSS                             = 3.0000 DB.
      LOWER PASSBAND EDGE FREQUENCY             = 8.8500000D+02 HZ.
      UPPER PASSBAND EDGE FREQUENCY             = 1.3000000D+03 HZ.
    SPECIFIED STOP BAND
      MULTIPLICITY OF ZERO AT ZERO             = 2
      MULTIPLICITY OF ZERO AT INFINITY         = 2
      OVERALL FILTER DEGREE                    = 4
      INPUT TERMINATION                        = 1.0000000D+00 OHMS
      OUTPUT TERMINATION                       = 0.0000000D+00 OHMS
      REQUESTED TERMINATION RATIO              = 0.0000000D+00

WISH TO SEE TRANSFER FUNCTION: Y/N
> N
ENTER THE NUMBER OF INCREMENTS (UP TO 5). NO ANALYSIS, ENTER: 0
> 0
REALIZATION - ACTIVE: A, PASSIVE: P, DIGITAL: D, NO SYNTHESIS: E
> A
ENTER SUBTITLE. IF NO MORE SYNTHESIS, ENTER: END
> AKTIVE REALISIERUNG MIT ZWEI STUFEN JE ZWEITER ORDNUNG
CASCADE. C, FOLLOW-THE-LEADER: F OR LEAPFROG: L
> C
WISH ALL BANDPASS BLOCKS: Y/N
> Y


AKTIVE REALISIERUNG MIT ZWEI STUFEN JE ZWEITER ORDNUNG

       RESULTS OF THE CASCADE SYNTHESIS

       BIQUADRATIC BLOCK COEFFICIENTS
              IN ASCENDING ORDER




             NUMERATOR            DENOMINATOR

SECTION NO.  1
    0.0000000D+00          3.3643451D+07
    7.3203664D+02          7.1555878D+02
    0.0000000D+00          1.0000000D+00

SECTION NO.  2
    0.0000000D+00          6.1318594D+07
    4.6556174D+03          9.6603118D+02
    0.0000000D+00          1.0000000D+00
```

```
COMMAND:                              COMMAND:

>  DES  1                             > DES

ENTER CAPACITANCE VALUE               ENTER SECTION NUMBER

>  10E-9                              > 2

                                      ENTER CAPACITANCE VALUE

                                      > 10E-9


SECTION NO. 1                         SECTION NO. 2


        .      .                            .      .

        .    R      1.77437E+05            .    R      2.78997E+04

        .      .                            .      .

    ....R........  7.52394E+03         ....R........  6.61384E+03

        .    .    .                         .    .    .

        .    C    . 1.00000E-08            .    C    . 1.00000E-08

        .    .  C  1.00000E-08            .    .  C  1.00000E-08

        .    .    .                         .    .    .

        .  ..R. .  4.11804E+04            .  ..R. .  3.05031E+04

        .    .    .                         .    .    .

    ..R..    .   . 2.98901E+03         ..R..    .   . 2.98901E+03

        .  ..     .                         .  ..     .

        .  ..   .                           .  ..   .

        . .\- +/  .                         . .\- +/  .

        . . \ /   .                         . . \ /   .

        . R  V   . 1.00000E+4             . R  V   . 1.00000E+04

        .  .   .   .                         .  .   .   .

        .  ..........                        .  ..........

        .      .                            .      .


SECTION PROVIDES 180 DEGRESS          SECTION PROVIDES 180 DEGREES
ADDITIONAL PHASE                      ADDITIONAL PHASE
```

Microwave lowpass filter

S/FILSYN in this case is being used for designing a lowpass filter with a cut-off frequency of 5 GHz. In the passband range, a maximum attenuation of 0.2 dB is to be maintained in the Chebyshev sense. The line sections used for realisation should possess a quarter of the wavelength at 10 GHz. Two $\lambda/4$-lines should appear in the ladder structure and multiplicity of three is selected for the zero at 10 GHz. The solution with S/FILSYN follows.

```
***** S/FILSYN *****

   RELEASE 1.0  VERSION 01    01/01/82

PLACER: P, FILSYN: F, LADDER: L, DIGITAL: D, ACTIVE: A OR END: E
> F
ENTER TITLE
> MIKROWELLEN-TIEFPASS
FILTER KIND - LUMPED: 0, DIGITAL: 1 IR MICROWAVE: 2
> 2
ENTER QUARTER WAVE FREQUENCY IN HZ
> 10E9
FILTER TYPE - LOWPASS: 1, HIGHPASS: 2 OR BANDPASS: 4
> 1
UPPER EDGE OF THE PASSBAND IN HZ
> 5E9
PASSBAND - MAX.-FLAT: 0, EQUAL-RIPPLE: 1, FUNCTIONAL INPUT:2
> 1
WHAT IS THE BAND EDGE LOSS IN DB
> 0.2
STOPBAND - MONOTONIC: 0, EQUAL-MINIMA: 1 OR SPECIFIED: 2
> 2
ENTER NUMBER OF UNIT ELEMENTS
> 2
ENTER MULTIPLICITY OF TRANSMISSION ZERO AT QUARTER-WAVE FREQUENCY
> 3
ENTER NUMBER OF FINITE TRANSMISSION ZEROS
> 0
ENTER INPUT TERMINATION IN OHMS
> 50
ENTER OUTPUT TERMINATION ( 0. INDICATES OPEN OR SHORT)
> 50

MIKROWELLEN-TIEFPASS
   LOW-PASS FILTER
      EQUAL RIPPLE PASS BAND
         BANDEDGE LOSS                  = 0.2000 DB.
         MAX. PASSBAND VSWR             = 1.5386
         UPPER PASSBAND EDGE FREQUENCY  = 5.0000000+09 HZ.
         QUARTER-WAVE FREQUENCY         = 1.0000000+10 HZ.
      SPECIFIED STOP BAND
         MULTIPLICITY OF ZERO AT QUARTER-
         WAVE FR.                       = 3
         NUMBER OF UNIT ELEMENTS        = 2
         NUMBER OF FINITE TRANSMISSION
         ZERO PAIRS                     = 0
         OVERALL FILTER DEGREE          = 5
         INPUT TERMINATION              = 5.0000000D+01 OHMS
         OUTPUT TERMINATION             = 5.0000000D+01 OHMS
         REQUESTED TERMINATION RATIO    = 1.0000000D+00
         NEAREST AVAILABLE TERMINATION
         RATIO                          = 1.0000000D+00

WISH TO SEE TRANSFER FUNCTION: Y/N
> N
ENTER THE NUMBER OF INCREMENTS (UP TO 5). NO ANALYSIS, ENTER: 0
> 0
EXISTING POLYNOMIALS ARE: ES OS    OD
ENTER SUBTITLE. NO MORE SYNTHESES: END
> Realisierung Mikrowellentiefpass
LATTICE: L, COMPUTER CONFIG.: C, INPUT SIDE: IN, OUTPUT SIDE: OUT
> C
```

Realisation of microwave low-pass filter

```
                    **** ALL VALUES ARE IMPEDANCES ****
```

```
    1  ....R....      5.0000000D+01        COMMAND:
       .       .                           > IB 3 6
       *       *                             3  ...C....      3.7944610D+01
    3  *  U E  *      2.6138410D+01           .       .
       *       *                              *       *
                                           5  *  U E  *      8.4007706D+01
       *       *                              *       *
    5  *  U E  *      7.6664142D+01
       *       *                          11  ...C....      2.4211666D+01
    7  ....C....      2.4899055D+01           .       .
                                              *       *
    8  .      L       6.5212859D+01       13  *  U E  *      8.4007706D+01
                                              *       *
    9  ....C....      3.7944610D+01
                                          17  ...C....      3.7944610D+01
   11  ....R....      5.0000000D+01
                                          19  ....R....      5.0000000D+01
  COMMAND:                                        .
  > IB 5 7                                COMMAND:
    6  .      L       5.7869295D+01        > SQU
       .       .
       *       *                          Realisierung Mikrowellentiefpass
    9  *  U E  *      1.8794847D+01       **** ALL VALUES ARE IMPEDANCES ****
       *       *                            1  ....R....      5.0000000D+01
                                              .       .
   12  .      L       6.5212859D+01        3  ....C....      3.7944610D+01
                                              .       .
   13  ....C....      3.7944610D+01           *       *
                                           5  *  U E  *      8.4007706D+01
   15  ....R....      5.0000000D+01           *       *
  COMMAND:
  > IB 9 12                                 7  ....C....      2.4211666D+01
    9  ....C....      2.4211666D+01           .       .
       .       .                              *       *
       *       *                           9  *  U E  *      8.4007706D+01
   11  *  U E  *      8.4007706D+01           *       *
       *       *
                                          11  ....C....      3.7944610D+01
   15  ....C....      3.7944610D+01           .       .
                                          13  ....R....      5.0000000D+01
   17  ....R....      5.0000000D+01
       .
```

Circuit file for SUPERCOMPACT:

```
        TY HGMWTP.DAT
    F1 :  1.00000E+10
    EL :   90.00
    BLK
    * RES   1  0  R = 5.00000E+01
      OST   1  0  ZO= 3.79446E+01  ELEN=EL  F=F1
      TRL   2  1  ZO= 8.40077E+01  ELEN=EL  F=F1
      OST   2  0  ZO= 2.42117E+01  ELEN=EL  F=F1
      TRL   3  2  ZO= 8.40077E+01  ELEN=EL  F=F1
      OST   3  0  ZO= 3.79446E+01  ELEN=EL  F=F1
    * RES   3  0  R = 5.00000E+01
    A:2POR  1  3
      END
    FREQ
      STEP  500 MHZ  9.5 GHZ  500 MHZ
    END
    OUT
    PRI  A  S
    END
```

Chapter 8

Receiver components and systems

A. Rupp

8.1 Introduction

HF receiver technology has undergone extremely rapid development over the past ten years, thanks to the technological progress in the semiconductor and line engineering sectors. In the frequency range up to 4 GHz, both the silicon transistor and the GaAs FET are used, while above 4 GHz the GaAs FET provides substantially better noise factors. The HEMT (high electron mobility transistor) is lower in noise in the high GHz range than the GaAs FET.

The stripline is used mainly in the frequency range from 300 MHz to about 30 GHz. It serves as a signal line and as a line element in HF circuits, and thus allows for a receiver to be created in integrated technology in the smallest possible space. Below 300 MHz, in some cases even as far as 1 GHz, HF circuits are still formed with lumped components. Miniaturisation of the circuits is achieved by using components such as transistors, resistors, and capacitors in chip form and applying these to the carrier material by means of thin-layer and thick-layer technology.

This state of the art approach is often referred to as hybrid technology. It is likewise used in circuits in the GHz range. Whole receiver modules have been created in the smallest possible space, such as HF input stages, IF amplifiers, and video amplifiers, in monolithic technology. The prospects are promising for substantial costs reductions in large production numbers (see Section 8.5).

The arguments in favour of miniaturising HF receivers are the smaller spatial requirement and substantially lower costs. Attractively priced receivers are a precondition for the construction of, for example, signal-processing antenna systems for tasks such as location, navigation, and telecommunications both in ground systems as well as on board aircraft and in satellites. Such a system consists of several thousands of individual transmitters and just as many receivers and possesses the capability, by means of a computer, of electronically controlling the antenna properties, such as multibeam reception, swinging the direction of the beam, optimisation of minor lobes, lobe width, gain, and masking sources of interference.

8.2 Receiver characteristics

The task of an HF receiver in the sense of telecommunications technology is to pick up electromagnetic signals, select them, and amplify them. In most cases, the amplifier must also transform these signals (e.g. by demodulation) into a form that is suitable for the processing which follows.

The capacity of a receiver to detect and amplify input signals is limited by noise, which partly passes to the receiver via the antenna as external noise, and is partly produced by the receiver itself. A good receiver is accordingly characterised by a low noise factor. Associated with this is a high level of sensitivity and a high signal-to-noise ratio. The receiver must also have high dynamic range, so as to be able on the one hand to detect the smallest signal, and, on the other, not to be saturated by the largest signal that is to be anticipated.

Other important features are the bandwidth, amplification, stability, mechanical robustness, and, last but not least, the receiver's reliability.

8.2.1 Receiver noise

The smallest signal that a receiver can process is limited by noise, which is essentially composed of three parts:

– antenna noise
– line noise
– the noise of the receiver itself.

By agreement, all noise fractions are defined as if they derived from thermal noise from an equivalent resistance at a temperature T. When matched, this supplies a noise power value of

$$N = kTB_n \qquad (8.1)$$

$$
\begin{aligned}
k &= \text{Boltzmann constant: } 1.38 \times 10^{-23} & &\text{(Ws/K)}\\
T &= \text{absolute temperature} & &\text{(K)}\\
B_n &= \text{noise bandwidth} & &\text{(Hz)}
\end{aligned}
$$

For $T = T_0 = 290$ K and $B_n = 1$ Hz, the following applies:

$$N = kT_0 = -174 \text{ dBm} \qquad (8.2)$$

8.2.1.1 Noise factor

The ideal receiver or amplifier delivers no additional noise; i.e. the signal-to-noise (S/N) ratio which pertains at the input will be transferred without change. As a measure of the quality of an amplifier, the noise factor F is introduced:

$$F = \frac{S_{in}/N_{in}}{S_{out}/N_{out}} = \frac{N_{out}}{kTB_n \cdot G} \qquad (8.3)$$

$$N_{out} = F \cdot kTB_n \cdot G \qquad (8.4)$$

$$G = \frac{S_{out}}{S_{in}} = \text{amplification}; \; N_{in} = kTB_n$$

As a reference temperature, the room temperature $T_0 = 290$ K is generally chosen. The noise factor F is in most cases given in dB, and is then designated as the noise factor NF.

$$NF = 10 \log F \; [dB]$$

N_{out} is composed of the amplified input noise GkT_0B_n and the additional noise N_z generated internally by the amplifier.

$$F = \frac{kT_0B_n \cdot G + N_z}{kT_0B_n \cdot G} = 1 + \frac{N_z}{kT_0B_n \cdot G} \qquad (8.5)$$

8.2.1.2 Noise temperature

Indicating the noise factor related to the temperature T_0 is impractical with amplifiers with very small noise factors. In these cases, the calculation is made rather with the effective noise temperature T_{eff}. We think of the noisy amplifier being replaced by one that is free of noise, at the input terminals of which the noise power of a resistor with the temperature T_{eff} is imposed, which after amplification produces the additional noise power N_z. The following then applies:

$$N_z = kT_{eff}B_nG \qquad (8.6)$$

Inserted in eqn. 8.5, the simple equation is derived:

$$F = 1 + \frac{T_{eff}}{T_0} \qquad (8.7)$$
$$T_{eff} = T_0(F - 1) \qquad (8.8)$$

8.2.1.3 Friisch formula

If two amplifiers are connected together, with the same noise bandwidth B_n but differing noise factors and available power amplification, the total noise factor F_{tot} can be calculated as follows. From eqn. 8.4,

$$\xrightarrow{\;kT_0B_n\;} \boxed{F_1, \, G_1, \, B_n} \xrightarrow{\hspace{3cm}} \boxed{F_2, \, G_2, \, B_n} \xrightarrow{\;N_{out}\;}$$

$$N_{out} = F_{tot}kT_0B_nG_1G_2$$

$$N_{out} = F_1kT_0B_nG_1G_2 + N_{z2}$$

Using eqns. 8.6 and 8.8,

$$F_{tot} = F1 + \frac{F2 - 1}{G_1} \tag{8.9}$$

In general:

$$F_{tot} = F1 + \frac{F2 - 1}{G_1} + \frac{F3 - 1}{G_1 G_2} + \cdots \frac{F_n - 1}{G_1 G_2 \ldots G_{n-1}} \tag{8.10}$$

By analogy,

$$T_{eff} = T_{eff_1} + \frac{T_{eff_2}}{G_1} + \frac{T_{eff_3}}{G_1 G_2} + \cdots \frac{T_{eff_n}}{G_1 G_2 \ldots G_{n-1}} \tag{8.11}$$

8.2.1.4 Noise from imperfect components

Imperfect components, such as filters, cables, or hollow conductors, have attenuation, which contributes to the total noise. The noise contribution is provided by the equivalent noise temperature or noise factor. The noise power kT_0B_n which is fed into the quadripole is attenuated by the loss factor L_R (≥ 1).

$$\xrightarrow{kT_0B_n} \boxed{T_R, L_R} \xrightarrow{N_{out}}$$

Subject to the precondition that there is thermal balance in the component, the noise output power can be calculated as

$$N_{out} = kT_0B_n = kT_0B_n \cdot \frac{1}{L_R} + kT_RB_n \cdot \frac{1}{L_R}$$

where T_R is the noise temperature related to the input of the component. Hence,

$$T_R = T_0 (L_R - 1) \tag{8.12}$$

Inserting this result into eqn. 8.8, provides the important conclusion that the noise factor of an imperfect component coincides with the loss factor:

$$F_R = L_R \tag{8.13}$$

8.2.2 Sensitivity

The effective system noise temperature T_S of a receiver is calculated from eqn. 8.11 to

$$T_A \longrightarrow \boxed{\text{Antenna}} \longrightarrow \boxed{\text{Line}} \longrightarrow \boxed{\text{Receiver}}$$

$$G = 1 \qquad\qquad T_R, L_R = \frac{1}{G_2} \qquad\qquad T_E, G_3, F$$

$$T_S = T_A + T_R + T_E L_R \tag{8.14}$$

Using eqns. 8.8 and 8.12 we obtain

$$T_S = T_A - T_0 + L_R F T_0 \tag{8.15}$$

The sensitivity of a receiver is defined as the input signal power S_{min}, which produces a signal-to-noise ration $S_{out}/N_{out} = 1$ at the output of the receiver. Accordingly,

$$S_{min} = kT_S B_n = kB_n (T_A - T_0 + L_R T_0 F) \tag{8.16}$$

If the noise is applied to the antenna at room temperature, $S_{min} = F \cdot kT_0 B_n \cdot L_R$ or

$$S_{min} \,[dB] = 174 \,[dB] + F \,[dB] + B_n \,[dB] + L_R \,[dB] \tag{8.17}$$

e.g. for $L_R = 2 \,dB$, $F = 4 \,dB$, $B_n = 1 \,MHz$, we have $S_{min} = -108 \,dBm$.

8.2.3 Dynamic range

This term is understood to mean the maximum level differences at the input of the receiver which the receiver is capable of processing without a component in the receiver chain being saturated. The dynamic range of an amplifier can be calculated if the maximum and minimum transferable signal levels are known. The latter is derived from the sensitivity of the amplifier in accordance with eqn. 8.17. The maximum transferable signal level is defined by the 1 dB compression point. This is the point on the amplification curve at which the actual amplification line deviates from the straight by 1 dB and then moves into saturation (Figure 8.1).

Receivers often have the task of processing several frequencies simultaneously, e.g. f_1 and f_2. Because of the nonlinearity at high input levels, intermodulation products are then produced (see also Section 8.4.6), which, in addition to the harmonic frequencies, also contain the sum and difference frequencies, which in part lie very close to the fundamental frequencies (Figure 8.2).

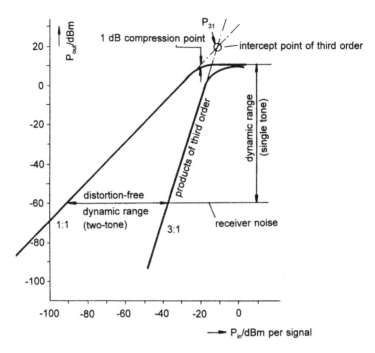

Figure 8.1   *Intercept diagram, dynamics*

Of particular significance in this case is the intercept point of the third order, which is specified by the manufacturer. If, as shown in Figure 8.1, this is inserted in the intercept diagram, which reproduces the dependency of the output power on the input power for signal frequencies (gradient 1:1) and the products of the third order (gradient 3:1) namely $2f_1 - f_2$ and $2f_2 - f_1$, then at a given level of the signal frequencies f_1 and f_2 the distortion-free dynamic range can be read off.

Example:
From Figure 8.1, with the values $G = 30\,dB$, $P_{31} = 20\,dBm$, $NF = 4\,dB$, $B = 100\,MHz$, $P_{1dB} = +10\,dBm$,

$$N_{out} = kToB \cdot G \cdot F = -174 + 80 + 30 + 4 = -60\,dBm$$

The dynamic range in the single-tone process is therefore 70 dB. The distortion-free dynamic range in the two-tone process can be read off from Figure 8.1, or determined in accordance with the following equation:

$$DR\,[dB] = 0.67\,(P_{31} - kT_0\,[dB] - G - 10\log B\,[MHz] - NF) = 53.6\,dB \quad (8.18)$$

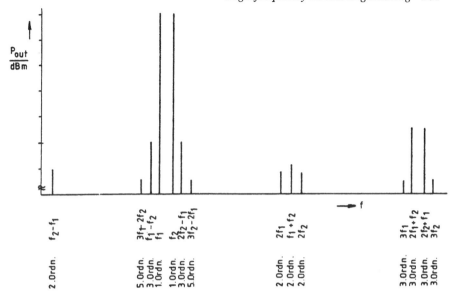

Figure 8.2   *Intermodulation products, input signals at f₁ and f₂*

The following example from radar technology shows dynamic values which can arise in practice for the receiver. From the radar equation, with constant transmitter power and constant antenna gain, the maximum and minimum receiver input level can be calculated as follows:

$$PE \sim \frac{\sigma}{R^4}$$

σ = back reflection cross-section
R = distance

$$PE_{max} \sim \frac{\sigma_{max}}{R^4_{min}} \qquad PE_{min} \sim \frac{\sigma_{min}}{R^4_{max}}$$

$$\text{Dynamic range } D = \frac{PE_{max}}{PE_{min}}$$

$$D \text{ [dB]} = 10 \log \frac{\sigma_{max}}{\sigma_{min}} + \frac{R_{max}}{R_{min}} \qquad\qquad (8.19)$$

$$= \text{target dynamic range} + \text{distance dynamic range}$$

Example:

σ_{max} =   1000 m^2 (ship, mountainside, house walls)

σ_{min} =   1 m^2 (boat, vehicle)

R_{max} =   32 km

R_{min} =   1 km

$D = 90$ dB $= 30$ dB $+ 60$ dB

There is no receiver that can process these dynamic ranges. Nevertheless, to be able to work with such a large level difference, most receivers are equipped with special circuits such as STC (sensitivity time control) and/or logarithmic amplifiers.

8.2.4   Calculation example

From Figure 8.8 the following are calculated:

−   Noise factor at the input of the HF amplifier, according to eqn. 8.10 is

$$F = F_{HFV} + \frac{F_{Fi} - 1}{G_{HFV}} + \frac{F_{Mi} - 1}{G_{HFV} \cdot G_{Fi}} + \frac{F_{IF} - 1}{G_{HFV} \cdot G_{Fi} \cdot G_{Mi}} + \ldots$$

$$= 1.58 + 0.0013 + 0.0136 + 0.0082 = 1.6 = 2.05 \text{ dB}$$

Thanks to its high amplification, the HF amplifier determines the total noise factor of the receiver practically alone,

−   losses from circulator to HF amplifier: 2.1 dB

−   sensitivity of the receiver according to eqn. 8.17 for a noise bandwidth of 10 MHz:

$$S_{min} = -174 + 2.05 + 70 + 2.1 = -100 \text{ dBm}$$

−   dynamic range of receiver: The largest signal is determined by the IF amplifier with 0 dBm at the output. Before the circulator, the following then applies: $S_{max} = -26.9$ dBm. From this, a dynamic of 73 dB is calculated at the IF amplifier output. The value is reduced by the following MTI limiting amplifier by 30 dB, since this is dimensioned in such a way that it moves into the limiting area even at 0 dBm, but still transfers the limited signal to the phase detector, free from phase shift. The phase detector can process the remaining dynamic range of 43 dB with no trouble at all.

8.3 Receiver concepts

8.3.1 Video detector receiver

The video detector receiver is the simplest representation of an HF receiver. Figure 8.3 shows the block circuit diagram of a wideband version. The main component is the detector, which converts the received HF energy into the video signal. The HF bandwidth usually lies in the gigahertz range, and is limited only by the bandwidth of the detector and the antenna. Frequency determination is not possible, because no selectivity is available. The sensitivity is indicated by the tangential signal sensitivity TSS. This is defined as the input signal that supplies a signal-to-noise ratio of 8 dB at the output of the video amplifier. The definition is derived from the observation of a pulse on the screen in the presence of noise energy (Figure 8.4).

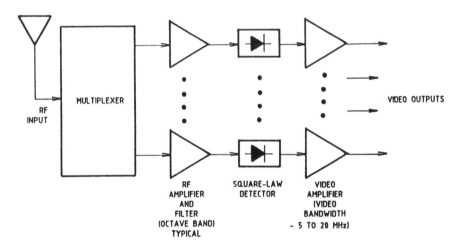

Figure 8.3   *Wideband video detector receiver*

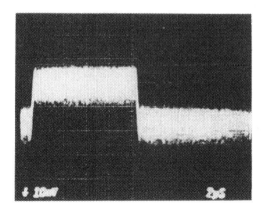

Figure 8.4   *Tangential signal sensitivity*

8.3.2 Superheterodyne receivers

Practically all receivers used today are superheterodyne receivers. In this arrangement, the HF signal coming from the antenna has an oscillator frequency superimposed on it in a mixer, and is converted into an intermediate frequency. A distinction is drawn between two types of heterodyne receivers, these being receivers with down converters, and receivers with up converters. Figure 8.5 shows a simplified block circuit diagram of a radio receiver as an example of a receiver with a down converter, The intermediate frequency is constant, and therefore independent of the frequency of the input signal. IF filters with very steep flanks can be created, which enables high selectivity to be attained. The frequency range of the oscillator determines the frequency range of the input signal.

The advantage with the use of the superheterodyne receiver in comparison with TRF receivers is the substantially higher sensitivity. The input signal f_{HF} is mixed in the mixer with the oscillator frequency f_0. The intermediate frequency f_{IF} then exists at the mixer output, in accordance with the formula:

$$f_{HF} = f_0 \pm f_{IF}$$

With a given oscillator frequency f_0, two HF input frequencies, located $2f_{IF}$ apart, supply the same intermediate frequency. The undesired input frequency (image frequency) is in most cases screened out before the mixer by means of a preselection filter or by a tunable filter (tuner). This method of image frequency suppression is conventional with narrowband systems.

Wideband systems require high intermediate frequencies, which must be greater than the frequency band which is to be transferred, for the image frequency band not to fall into the fundamental frequency band. At high frequencies, however, filters cannot be created with small bandwidths and steep flanks at acceptable costs, which means, for example, that frequency measurement with high resolution is an impossibility. This requirement can be obtained by repeated mixing; as an example, Figure 8.6 shows a receiver with double downconverter.

Modern wideband systems are increasingly making use of receivers with up converters. In this case, the input signal is converted by mixing with the oscillator to an intermediate frequency which is higher than the input frequency. The components required for this in the GHz range are now available; Figure 8.7 shows an example of this kind of wideband receiver.

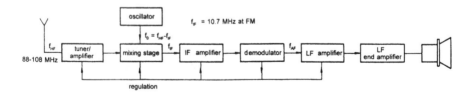

Figure 8.5   *Superheterodyne radio receiver (down converter)*

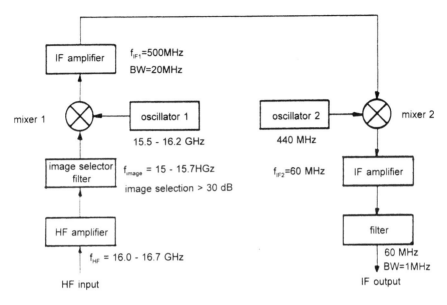

Figure 8.6   *Double superheterodyne receiver*

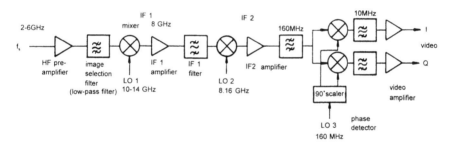

Figure 8.7   *Wideband receiver with image frequency suppression*

The input signal is converted from the range 2–6 GHz in mixer 1 by superimposition of the tunable LO_1 to the IF_1 of 8 GHz, with LO_2 to 160 MHz, and finally mixed to video with LO_3. The image frequency range of the system is between 18 and 22 GHz ($f_0 + f_{IF}$). The advantage of this receiver is that a simple lowpass filter can be used as the image selection filter, and the tuning range of LO_1 is smaller than one octave. A disadvantage is the need for three oscillators, two of which, however, are of fixed frequency.

8.3.3 Fully coherent radar receivers

A large number of situations arise in radar technology involving the task of differentiating between moving targets (such as aircraft) and fixed targets (such as ground reflections and reflections from buildings). Fixed targets do not provide any Doppler information and are suppressed in an appropriate circuit called MTI (moving target indication). Suppression is effected in a phase detector in which the sinusoidal oscillations of the COHO (coherent oscillator) are compared from pulse to pulse with those of the echo signal. Fixed targets provide constant video voltages at the output of the phase detector, while for moving targets temporally-variable video voltages exist because of the differing phase. Separation can be effected by means of an MTI filter.

In the fully coherent radar transmitter/receiver, Figure 8.8, the transmission frequency is created by the up-conversion of the STALO frequency (stabilised local oscillator) with the likewise quartz-stabilised COHO frequency. It is then pulsed with the aid of a PIN diode modulator, amplified to transmitting power from the transmitter stage by means of TWT, klystron, or cross-field amplifier, and radiated via the antenna.

The received signal passes via the transmitter/receiver separating filter, which in most cases is a circulator, to the TR limiter, the task of which is to protect the receiver against destruction by the transmission pulses or external transmission pulses. The HF-STC (sensitivity time control) which is connected is a time-dependent controlled attenuation element. It narrows the dynamic range of the receiver in such a way that targets of equal size at short range (large input signal) and long range (small input signal) are received and displayed with the same signal strength (distance-dependent gain control). The subsequent amplification of the input signal up to about 20 GHz is achieved nowadays exclusively with low-noise HF transistors. Mixing takes place, mostly to 60 or 160 MHz, after the preselection filter, as does amplification.

The MTI limiting amplifier connected transfers the input signal in phase to the phase detector already mentioned; the video output signal from this undergoes further processing in the following module.

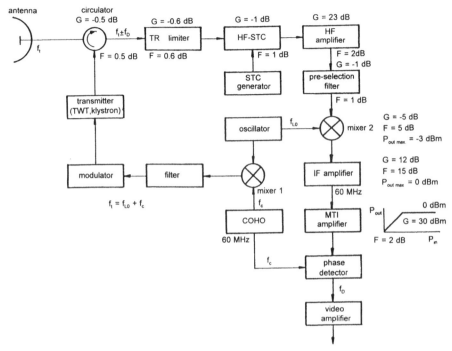

Figure 8.8 *Fully coherent radar transmitter/receiver*

8.3.4 Monopulse receivers

Figure 8.9 shows another example of a receiver, as used in secondary radar technology and in direction-finding engineering for azimuth direction determination. It consists essentially of two receiving channels, the outputs of which represent the amplified signals, mixed to IF, from two antenna images which are compared in the monopulse detector. The signal received from the antenna can, for example, be the response signal from a transponder on board an aircraft, or derived from a direction-finding transmitter.

A distinction is drawn between an amplitude and phase monopulse system. With the former, the amplitudes of two or more antenna signals are compared with one another. The antenna consists in most cases of a reflector and several exciters, connected together in a comparator of such a type that the sum (Σ) and difference (Δ) images are obtained, the output signals of which are evaluated in the receiver. In the event of the flying target deviating from the axis direction of the antenna, a video voltage will exist at the output of the monopulse detector, the size of which represents a value for the angle of deviation of the target. Since the phase angle is reversed by 180° during the passage of the target from one side of the antenna axis to the other, it is possible, by means of operational sign recognition in a phase detector, to

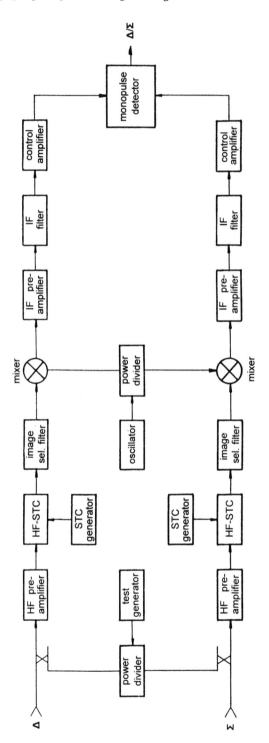

Figure 8.9 *Monopulse receiver*

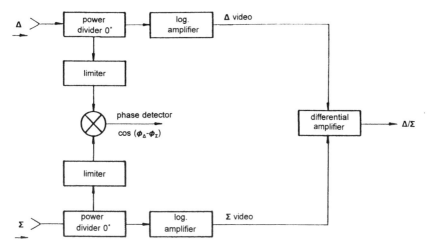

Figure 8.9a *Detector for amplitude monopulse*

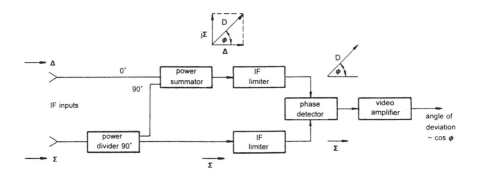

Figure 8.9b *Detector for phase monopulse*

determine the lateral deviation to the left or right (see Figure 8.9). The angle information acquired in this way serves either the purpose of azimuth determination or is used to track the antenna onto the target. The logarithmic amplifier shown in Figure 8.9a is used to restrict the dynamics of the receiver.

In the case of the phase monopulse process, the antenna consists of two or more identical individual antennas, the output signal amplitudes of which are practically the same size, but the phases of which differ, depending on the angle of incidence of the reception signal. Signals in the direction of the parallel antenna axes have the same phase. Figure 8.9b shows how the angle of deviation can be measured by means of a suitable circuit in the detector. The IF limiting amplifier indicated must, in this case, transfer the signal in phase.

Both monopulse receivers have strict requirements for synchronous operation of the two reception channels with regard to amplitude and phase, respectively. The block circuit diagram in Figure 8.9 accordingly contains a test generator and a control amplifier in each reception channel which enables amplification fluctuations of both channels to be corrected.

8.3.5 Wideband receivers

Military reconnaissance systems require wideband receivers, the task of which is, among other things, to determine the frequency of the input signal in the frequency range from 2–18 GHz and above, with a degree of precision of several MHz. In Figure 8.10 this task is resolved with the aid of a multichannel wideband superheterodyne receiver with identical downstream IFM (instantaneous frequency measurement) modules. This receiver is referred to as a channelised receiver. The entire frequency range to be monitored is subdivided into bands 2 GHz wide and converted into the baseband of 2–4 GHz. The precise frequency determination is carried out in the IFM receiver that follows. The input signal is divided at that point into two part signals, one of which is delayed by a known running time τ (Figure 8.11); this results in a phase displacement between the two part signals of $\theta = \tau$.

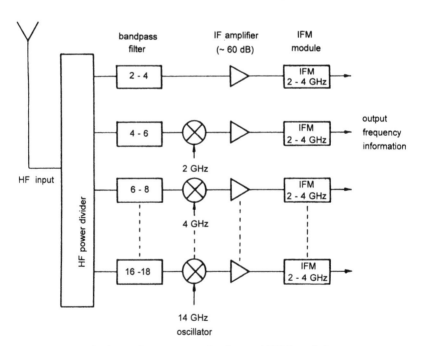

Figure 8.10 *Multichannel receiver with identical IFM modules*

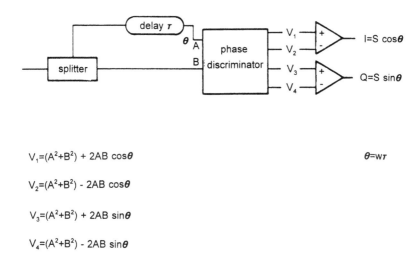

$V_1=(A^2+B^2) + 2AB \cos\theta$

$V_2=(A^2+B^2) - 2AB \cos\theta$

$V_3=(A^2+B^2) + 2AB \sin\theta$

$V_4=(A^2+B^2) - 2AB \sin\theta$

$\theta=w\tau$

Figure 8.11   *Principle of frequency measurement with IFM receiver*

Both part signals are now passed to a phase discriminator, also known as a correlator. This consists in principle of a 90° and a 180° hybrid, which serve as input port networks, and two downstream 90° hybrids, the outputs of which are connected to four detector diodes (Figure 8.12). The output voltages of the four detectors depend on the amplitude and phase of the input signals at the correlator.

At the output port network of the connected differential amplifier the signals I and Q occur, which are a function of the frequency which is to be determined. These

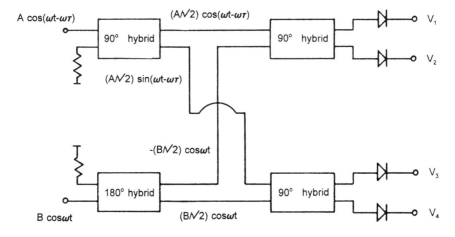

Figure 8.12   *Example of phase discriminator*

output signals usually undergo further processing in a digital processor. The delay
line is theoretically free of faults, but the correlator is not; errors are typically of ± 6°.
For a system of 2–4 GHz, this means a degree of precision of frequency measurement
of ± 34 MHz. If even greater precision is required, several correlators are connected
in parallel in the appropriate manner. An IFM system with four correlators and a
10-bit processor supplies a frequency resolution of approximately 2 MHz in the
frequency range from 2–4 GHz.

8.4 HF components

8.4.1 Low-noise transistor amplifiers

GaAs field-effect transistors, up to about 20 GHz, and also HEMTs individually
designed up to about 40 GHz, allow for HF amplifiers to be constructed with very
small noise factors. Figure 8.13 shows the noise factor of HF amplifiers from
manufacturers' catalogues. The 1 dB compression point is located at about + 10 dBm.

8.4.2 Logarithmic amplifiers

The logarithmic amplifier is a component that produces a video voltage at its output
which is proportional to the logarithm of the HF power present at the input:
$U = K \log P_{in}$. Accordingly, reception of signals with very high dynamic range is
possible, without limitation or saturation of the amplifier. An 80 dB dynamic range,
of input power from –80 dBm to 0 dBm, is typical. The rise in transfer function is
25 mV/dB, so that at an input level of 0 dBm a video voltage of about 2 V obtains.
The logarithmic amplifier is used in systems in which input signals are intended to

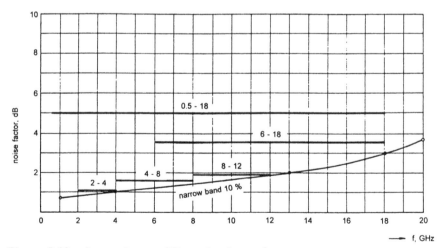

Figure 8.13 *Low-noise amplifier noise figure data*

be transferred simultaneously at very different amplitudes, and/or if it is intended that the high input dynamic range is to be restricted.

The logarithmic amplifier is used predominantly in the IF range. Because of the nonlinear behaviour, electrical data, such as dynamic range, bandwidth, and pulse response are changed during operation; this is to be taken into account when establishing the parameters of the receiver channel.

8.4.3 90° hybrid

The 90° hybrid provides a 3 dB power divider, and is of great importance in receiver technology. Figures 8.14a and b show schematic layouts of two possible design forms. When arms 2, 3 and 4 are terminated with $R = Z_0$, the input impedance at arm 1 likewise provides the value $R = Z_0$. The connection point 3 is disconnected from the input 1, and vice versa. Half the power of 1 and 3 is conducted to the connections 2 and 4, respectively. An important factor is that, with the 90° hybrid, a phase difference of 90° exists between the outputs 2 and 4.

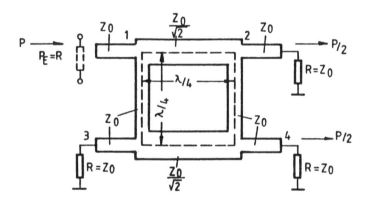

Figure 8.14a *90° hybrid (2dB divider)*

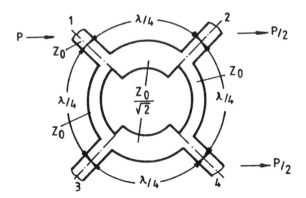

Figure 8.14b *90° ring hybrid*

Below 1 GHz, the 90° hybrid can also be set up with inductances and capacitances, as Figure 8.15 shows. This divider is also often designated as a quadrature coupler because of the 90° phase difference at the two outputs. The following dimensioning arrangement of the components is derived:

$$C_1 = 1/(Z_0\omega_0 \sqrt{K}) \tag{8.20}$$

$$\omega_0 2L (C_0 - C_1) = 1 \tag{8.21}$$

$$L = 2C_0Z_02/(1 + \omega_02C_02Z_02) \tag{8.22}$$

K = proportion of output power at ports 2 and 4
$K = 1$, with the 3 dB divider

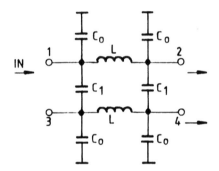

Figure 8.15   *90° hybrid*

8.4.4 *Wilkinson dividers (0° divider)*

The Wilkinson divider (Figure 8.16) is likewise a 3 dB power divider, but in comparison with the ring hybrid there is no phase difference at outputs 2 and 4. The wave impedance of the outputs 2 and 4 are transformed via $\lambda/4$ long lines of 50 to 100 Ω, so that, by means of a parallel connection at input 1, 50 Ω again occurs. The Wilkinson divider is then matched. In addition to this, outputs 2 and 4 are decoupled via the two $\lambda/4$ bypass lines and the internal terminating resistor. With ports 2 and 4 equally loaded, the stabilising resistor consumes no energy. The bandwidth of the divider is about 25% and decoupling is at about 20 dB. Figure 8.17 is an example of a two-stage power divider, which is therefore wideband.

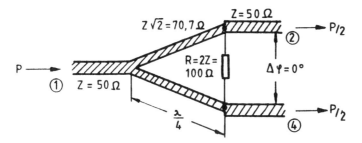

Figure 8.16a *Wilkinson divider in normal design format*

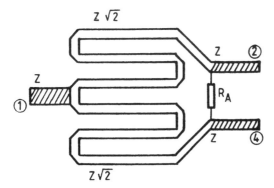

Figure 8.16b *Wilkinson divider in meander form*

Figure 8.17 *Two-stage Wilkinson divider*

8.4.5 Hybrid transformers

This kind of transformer is formed from lumped components and is used below 1 GHz. It is comparable in effect with the Wilkinson divider, and has the same properties (Figure 8.18):

— The input signal at port D is divided into two output signals, which are equal in amplitude and phase.

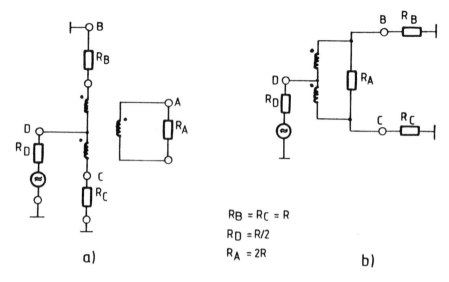

Figure 8.18   *Hybrid transformer*

(a) Ideal
(b) Simplified

— In a 50 Ω system, R = 50 Ω, and the input impedance at port D = 25 Ω. A matching transformer is therefore required for converting 25 to 50 Ω.

— The outputs B and C are decoupled from each other. At input port B, the signal undergoes a phase displacement of 180° by means of the transformer, but 0° via the resistance R_A. Because the impedance of the transformer is 100 Ω, and the internal impedance is likewise 100 Ω, the signals incoming at gate C are of equal amplitude, and cancel one another out.

— In the ideal situation (symmetry and matching), no current flows via R_A; conversely, no signal passes for infeed at A to port D.

In general, at any random infeed into one of the four ports, it is always the case that the opposite port is decoupled, and the power division takes place onto the two remaining ports. The hybrid accordingly has the effect of a magic T, i.e. infeed to port C: distribution to A and D; B is decoupled.

8.4.6 Receiver mixers

The superheterodyne process has proved its value for the reception of high-frequency signals. According to the principle, the HF signal of frequency ω_S is overlaid by the local oscillator (LO) on the frequency ω_0 in the mixer, and converted up to the frequency $\omega_{IF} = \omega_S - \omega_0$. In most cases, Schottky barrier diodes are used as mixer diodes, which are available in a variety of designs. Low-barrier diodes are used for low oscillator power values (approximately 0 dBm), and medium-barrier diodes for large oscillator powers (20 dBm). The greater the oscillator power, the higher the intercept point and the 1 dB compression point, and therefore the greater the dynamic range. Other mixers use bipolar or field-effect transistors.

The mixer diode is a nonlinear switching element (characteristic line), for which the relationship between output current and imposed voltage can be determined mathematically by means of a Taylor series:

$$i = C_0 + C_1 \cdot u + C_2 \cdot u^2 + C_3 \cdot u^3 + ... \tag{8.23}$$

If two sinusoidal voltages of different frequencies are imposed on the diode,

$$u = U_0 \cos \omega_0 t + U_s \cos \omega_s t \tag{8.24}$$

then a signal appears at the output of the mixer of the following form (see also Figure 8.2):

$$i = C_0 + 1/2\ C_2\ U_s^2 + 1/2\ C_2\ U_0^2 ... \qquad \text{direct components}$$

$$+ C_1\ U_s \cos \omega_s t + C_1\ U_0 \cos \omega_0 t ... \qquad \text{amplified useful signal}$$

$$+ (3/4\ C_3\ U_s^3 + 3/2\ C_3\ U_s U_0^2 ...) \cos \omega_s t$$

$$\text{first order}$$

$$+ (3/4\ C_3\ U_0^3 + 3/2\ C_3\ U_0 U_s^2 ...) \cos \omega_0 t$$

$$+ (1/2\ C_2\ U_s^2 + ...) \cos 2 \omega_s t \qquad \text{first harmonic of } \omega_s$$

$$+ (1/2\ C_2\ U_0^2 + ...) \cos 2 \omega_0 t \qquad \text{first harmonic of } \omega_0 \tag{8.25}$$

$$+ (1/4\ C_3\ U_s^2 + ...) \cos 3 \omega_s t \qquad \text{second harmonic of } \omega_s$$

$$+ (1/4\ C_3\ U_0^2 + ...) \cos 3 \omega_0 t \qquad \text{second harmonic of } \omega_0$$

$$+ (C_2 U_s U_0 + ...) \cos (\omega_0 - \omega_s) t \qquad \text{second order differential frequency}$$

$$+ (C_2 U_s U_0 + ...) \cos (\omega_0 + \omega_s) t \qquad \text{second order summation frequency}$$

$$+ (3/4\ C_3 U_s^2 U_0 ...) \cos (2 \omega_s - \omega_0)t$$

$$\text{third order differential frequency}$$

$$+ (3/4\ C_3 U_s U_0^2 ...) \cos (2 \omega_0 - \omega_s)t$$

$$+ (3/4\ C_3 U_s^2 U_0 ...) \cos (2 \omega_s + \omega_0)t$$

$$\text{third order summation frequency}$$

$$+ (3/4\ C_3 U_s U_0^2 ...) \cos (2 \omega_0 - \omega_s)t$$

$$+ ...$$

As seen, the new frequencies conform to the law

$$\omega_z = n\omega_s \pm m\omega_0 \text{ with } z = |n| + |m| \tag{8.26}$$

In addition to the desired intermediate frequency, $\omega_{IF} = |\omega_0 - \omega_s|$, additional frequencies occur, which can in turn be mixed in the diode (Figure 8.19). An important factor in this is a new frequency $2\omega_0 - \omega_s = \omega_i$, the image frequency, which when remixed with ω_0 again produces the IF, and is superimposed on the

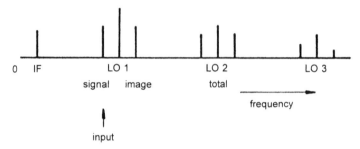

frequency spectrum of a mixer

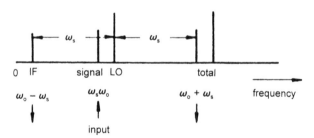

frequency spectrum created by signal and oscillator ground wave

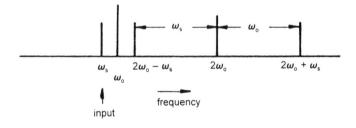

frequency spectrum created by signal and oscillator harmonics

Figure 8.19   *Frequency spectrum of mixer*

original IF. The conventional mixer is accordingly unable to differentiate between input signals ω_s and the image frequency ω_i.

If the mixer is upstream of an HF amplifier, it is important that the amplified image band noise does not reach the mixer, so as not to make the noise factor of the mixer any worse. This noise is suppressed by an image selection filter before the mixer or by what is referred to as the image selection mixer (Section 8.4.8).

The term 'conversion loss' is applied to the relationship between the power available at the signal frequency to the power available at the intermediate frequency. The quality of a mixer is designated by its noise factor, which is more or less the same as the mixing loss. Depending on the bandwidth, the noise factors of the mixer are nowadays between 4 and 8 dB.

Diode mixers can in general be divided into five groups in accordance with their design forms: single-ended mixers, single-balanced mixers, double-balanced mixers, image rejection mixers, and image enhanced mixers.

Figure 8.20 shows a two-diode mixer in stripline technology, with an IF pre-amplifier connected downstream. This operates in the frequency range from 1.2–1.4 GHz, at an output IF of 30 MHz.

Figure 8.20   *Mixer in stripline technology*

8.4.7 Ring mixer and applications

Below 1 GHz, use is frequently made of what is known as a ring mixer, also referred to as a double balanced mixer, as a compact module; Figure 8.21 shows how this works. One of the outstanding features of the ring mixer is its excellent isolation between the three inputs, HF, LO and IF. Typical isolation values of a ring mixer of 0.5–500 MHz in the upper frequency are about 35 dB, in the intermediate range 40 dB, and 50 dB in the lower range. If CR_1 and CR_2 and the LO transformer are symmetrical, then the same voltage exists at point A as at the middle tapping point of the transformer, in other words ground potential.

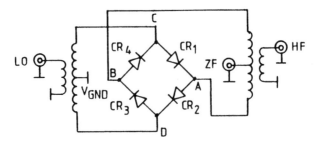

Figure 8.21 *Ring mixer*

If CR_3 and CR_4 are symmetrical, then the same applies to point B. This means that there is no voltage difference between A and B, and therefore no voltage either at the HF or IF connections. Considering the voltage conditions of the HF input, the same decoupling can be determined at the LO and IF connections.

Another important property of the ring mixer is the fact that the IF connection is DC coupled to the diodes. One can therefore deduce the following properties for the ring mixer:

— The ring mixer can be used in wideband applications.

— It functions as a phase detector if signals of the same frequency are imposed at the HF and LO connections. Substitute for eqn. 8.24:

$$u = U_o \cos (\omega t - \varphi_o) + U_s \cos (\omega t - \varphi_s)$$

and from eqn. 8.25 we obtain:

$$i = i_o + K \cos (\varphi_o - \varphi_s) + \text{components of higher frequencies}$$

which are suppressed by filtering. The phase detector accordingly supplies at its output a direct voltage which is proportional to the phase difference of the two HF signals.

— It serves as a biphase modulator if the modulation signal is fed in at the LO port with low level (– 10 dBm is typical) and at the IF port (Figure 8.22). If the modulation signal selected is sufficiently large (typically + 7 dBm), then the diodes function as individual switches; i.e. at each zero crossing of the IF modulation signal the phase rotates by 180°. At the HF output the two sidebands then contain (HF + IF) and (HF – IF). The carrier and undesirably high sidebands are in this case suppressed (typically carrier suppression is 40 dB).

— If DC is imposed at the IF input, then an imbalance can be artificially created, with the result that the isolation between HF and IF is caused to deteriorate, to about 3 dB. This means that the ring mixer can also be used as an electronically controlled switch or pulse modulator, or even as an attenuator. This is of particular interest at lower frequencies in cases in which the PIN diode loses its characteristics as a variable resistance (for typical attenuation curve see Figure 8.23). The excitation currents can in this context be positive or negative. If the ring mixer is used as a switch or pulse modulator, the actuation pulse is located typically between 0 and 20 mA. The switching speed is less than 1 ns.

— The ring mixer can also be used as a current-controlled limiter. Typical transfer characteristics are given in Figure 8.24.

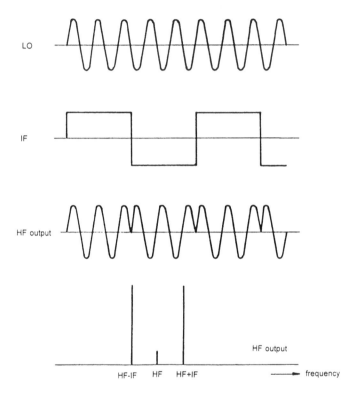

Figure 8.22   *Biphase modulator*

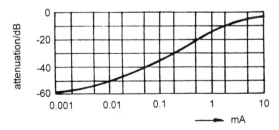

Figure 8.23   *Ring mixer as attenuator (typical attenuation curve)*

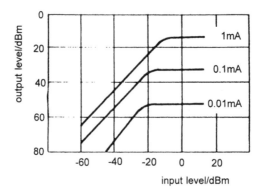

Figure 8.24   *Ring mixer as limiter (typical curves)*

8.4.8 Image rejection mixer (IRM)

This mixer is used predominantly in cases in which both sidebands around $\omega_0 - \omega_s$ and $\omega_0 + \omega_s$ are located so close to one another that they can no longer be separated by means of bandpass filters. Figure 8.25 shows how they function. It consists of a 0° power divider, two mixers and two 90° hybrids. Because of the known phase properties of the power dividers used, it can be seen from Figure 8.25 that division between desired frequency and image frequency can be effected by means of this circuit. The suppression of the image frequency signal in practice is about 25 dB, and depends on the amplitude and phase errors of the hybrids used.

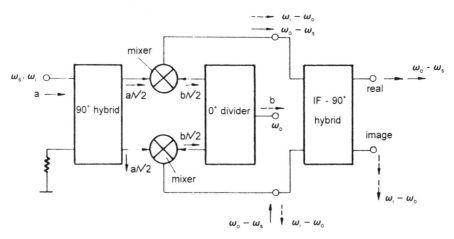

Figure 8.25   *Image selection mixer*

This value is sufficient not to impair the overall noise factor of the receiver, as is shown by the following explanation. Given is a receiver with HF preamplifier, image frequency selection, mixer and IF amplifier. According to eqn. 8.3, the following then applies to the noise factor at the input of the HF amplifier:

$$F' = \frac{N_{out} + \alpha N_{out}}{N_{in} \cdot G} = F(1 + \alpha)$$

α is the noise part of the image frequency. With image frequency suppression $I[dB] = -10 \log \alpha$,

$$F' = F(1 + 10^{-I/10})$$

The IRM is in general incorporated in stripline technology, and below 1 GHz it is even offered by a number of companies as an integrated module, with the transformer, capacitors, and diodes being located in a housing with connection pins for soldering into a conductor path.

8.4.9 HF attenuator

Receiver technology frequently requires continuously or digitally changeable, electrically controlled attenuators for automatic gain control, for adjusting and monitoring the signal level, and for amplitude demodulation. While the principle of the bridged T-element (Figure 8.26) is applied up to about 500 MHz, above this circuits predominate with 0° and 90° power dividers, most of which are designed in stripline technology (Figure 8.27 and 8.28). PIN diodes are used as variable resistors. If the size of the control current imposed is altered, the resistance of the PIN diode pair is also changed and therefore so is the power reflected at the diodes.

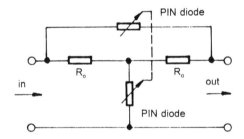

Figure 8.26 *Attenuator with bridged T*

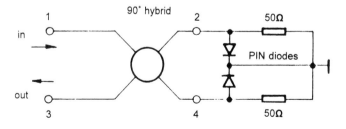

Figure 8.27 *Attenuator with 90° hybrid*

In the circuit shown in Figure 8.27, because of the 90° phase differential at ports 2 and 4, the entire reflected power (at the same high reflectance factor of the diode pair) passes to the output port 3. The attenuator is matched. Because of the arrangement of the diodes in Figure 2.28, displaced by λ/4, a displacement of 180° occurs, with the result that the reflected power is absorbed in the matching resistor. An attenuator implemented in the L-band with three diode pairs supplies about 80 dB maximum attenuation and a minimum attenuation of 0.8 dB (Figure 8.29). The control electronics are designed in such a way that a 'dB-linear' connection exists between the attenuation and the imposed voltage, e.g. 10 dB/V.

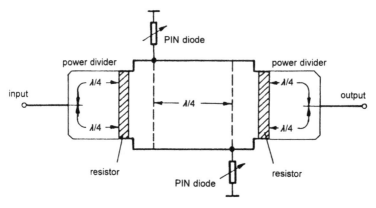

Figure 8.28   *Attenuator with 0° divider*

8.4.10 Phase shifter

The simplest type of phase shifter is created by changing the line length, for example in stripline arrangements. In Figure 8.30, the branch point A represents either a solder connection for one-off matching of the circuit, or A is a PIN diode switch, so that the example represents a three-bit phase shifter. Instead of the PIN diodes, high-isolating transistors can also be used, such as GaAs FETs.

Another type of phase shifter makes use of the properties of the 90° hybrid, with changeable short-circuit lines to ports 2 and 4. The circuit in Figure 8.31a is used mainly below 500 MHz. It consists in principle of a hybrid transformer and two LC elements. With the aid of the ganged capacitors, the length of the short-circuit line, and thus of the phase, can be manually adjusted. If the capacitors are replaced by varactor diodes, one obtains a voltage-controlled phase shifter (Figure 8.31b). The phase can be continuously adjusted.

Figure 8.31c shows the principal circuit diagram of a digitally controlled phase divider, in which the line is switched by the length l. The switching elements used are almost exclusively PIN diodes. These types of phase shifters are implemented in most cases in stripline technology, and are used a great deal with phase-controlled antenna systems.

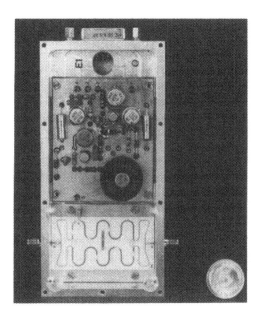

Figure 8.29 *Attenuator*

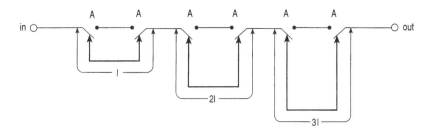

Figure 8.30 *Three-bit phase shifter with bypass line*

A third type of phase shifter makes use of the line theory, and is accordingly referred to as the loaded line phase shifter. At intervals of $\lambda/4$ along a line, susceptances of the same number, implemented by PIN diodes, are switched from the capacitive to the inductive state. The phase displacement depends on the size of the susceptance.

The hybrid-coupled phase shifter requires the smallest number of diodes for a specified number of bits, and is accordingly the most used. For a 4-bit phase shifter (smallest phase increment 22.5°), 30 diodes are needed for the line phase shifter, 16 for the phase shifter with bypass line, and eight PIN diodes for the hybrid-coupled phase shifter.

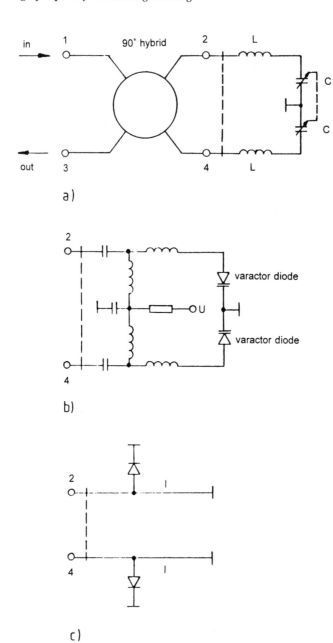

Figure 8.31 *Hybrid-coupled phase shifter*

(a) LC short-circuit shifter
(b) Voltage-controlled phase shifter
(c) Digitally controllable phase shifter

8.5 MMIC technology

The task of 12 GHz satellite receivers is to convert the 12 GHz signal at the reception point to the frequency range of the already existing radio and television receivers. The converter consists essentially of a low-noise preamplifier, a local oscillator, a mixer, and an IF amplifier.

Work is under way all over the world on the development of active phased-controlled antenna arrays for communications and radar tasks on board aircraft/satellites and on the ground. These antennas consist of a large number of individual emitters (several thousand), with transmission/reception modules downstream. Modules of this type contain elements such as switches, attenuators, phase shifters, low-noise preamplifiers, and power amplifiers.

These two examples show that future systems will be coming onto the market with a very large number of receivers, the creation of which will only be possible if it proves feasible to manufacture large numbers at acceptable costs. One possible way of doing this is the monolithic integrated microwave circuit (MMIC) technique, with GaAs as the substrate material. This is based on the integration of passive *and* active construction elements on the same substrate. The manufacture of MMICs can therefore be regarded as an extended manufacturing process in comparison with GaAs FET. In comparison with other technologies, the GaAs MMIC technology has the following advantages:
– compact design format
– low weight and volume

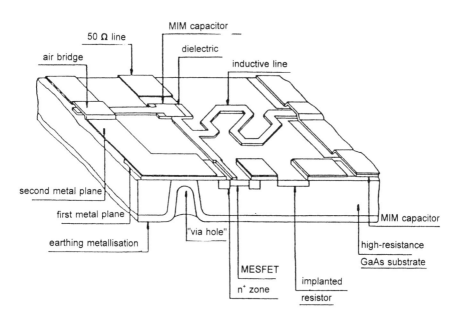

Figure 8.32 *MMIC chip technology*

– low production and integration costs
– increased reproducibility of the technical data on the basis of the uniform process
– higher reliability thanks to high integration (fewer wire connections)
– greater bandwidth.

Disadvantages that still persist today are the relatively high development costs and the still inadequate number of circuits on one wafer.

8.5.1 Typical MMIC manufacturing process

Figure 8.32 shows a technological example of an MMIC chip, on which have been located a MESFET, 50 Ω lines, an inductive line, MIM capacitors, resistors, metal bridges (air bridges), and plated-through contacts to a high-resistance GaAs substrate. The manufacturing process takes place in several stages; a typical layer formation is the following (Figure 8.33).

(i)   *Ion implantation*: A n-doped layer is implanted into a thick, semiconductive GaAs substrate (e.g. 400 µm thick), which forms the conductive path for the MESFET, and which also functions as an implanted resistance layer. The n-layer is followed by an n^+ layer, which is low resistance, and therefore serves to create the contact sections.

(ii)   *Insulation*: Components are insulated from one another by removal of the n layer with etches as far as the GaAs substrate.

(iii)   *Ohmic contacts*: A gold-germanium-nickel alloy is applied to the n^+ layer at high temperature (typically 400°C) in order to create the ohmic contacts.

(iv)   *Gate metallisation*: This consists of three layers, namely titanium, platinum, and gold. The first metallising layer is simultaneously applied by means of a Ti–Au alloy. This is used to strengthen the ohmic contacts, to form the passive elements, and as a subelectrode for the MIM capacitors (metal–insulator–metal).

(v)   *Dielectric*: A silicon nitride layer serves as passivation for the n-doped sectors (MESFET and resistance) and as a dielectric for the MIM capacitors.

(vi) 2nd *Metallising layer and air bridges*: This metallising provides the upper electrode for the capacitors and the connection between the components via the air bridge. In this context, layer thicknesses of 3–5 µm are created.

(vii)   *Plating-through*: Once the process stages described have been concluded, the back of the wafer is lapped to a thickness of 100–200 µm, plating-through (via holes) is carried out by means of a chemical caustic removal process, and the rear is metallised.

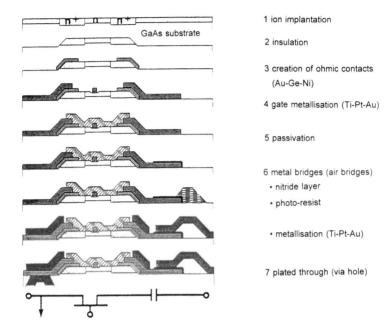

	1 ion implantation
	2 insulation
	3 creation of ohmic contacts (Au-Ge-Ni)
	4 gate metallisation (Ti-Pt-Au)
	5 passivation
	6 metal bridges (air bridges)
	• nitride layer
	• photo-resist
	• metallisation (Ti-Pt-Au)
	7 plated through (via hole)

Figure 8.33 *MMIC manufacturing process*

8.5.2 MMIC circuit elements

The most important component in an MMIC circuit is the transistor, in most cases implemented as a MESFET in GaAs. In view of the fact that the gate length of modern MESFETs is only 0.3–1 μm, there are major demands placed on the manufacturing process with regard to precision and reproducibility; 0.5 μm technology is available from a number of companies. Figure 8.34 shows a four-finger MESFET with gate length 1 μm, manufactured by AEG-TEG.

The semiconductor material GaAs has a dielectric constant of $\varepsilon_r = 12.6$ and a $\tan\delta = 16 \times 10^{-4}$, measured at 10 GHz. For a 50 Ω microstripline, a line width of 72 μm and losses of approximately 0.3 dB/cm are derived from this, with a substrate thickness $h = 100$ μm. This means that, with GaAs as substrate material, good microstriplines can be created for connecting MMIC components.

Because of the shortening factor of 0.36, this stripline is only to be used above 30 GHz as a component ($\lambda/4$ line, spur line); below this, it occupies too much space (one $\lambda/4$ long spur line at 10 GHz has a length of 2.7 mm). Accordingly, below 30 GHz, passive components are used almost exclusively, in lumped form, resistances in the form of epitaxy layers, capacitances in the form of gap-coupled lines, interdigital capacitors (up to 1 pF) or metal–insulator–metal capacitors (MIM, capacitance 330 pF/mm^2), inductances in the form of meander lines or, in most cases, in the form of spiral inductances with air bridges (Figure 8.35).

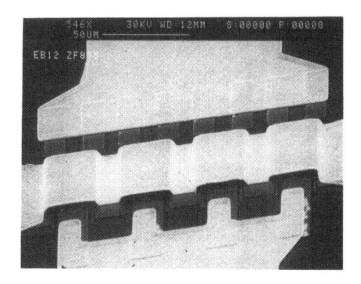

Figure 8.34   *GaAs MESFET*

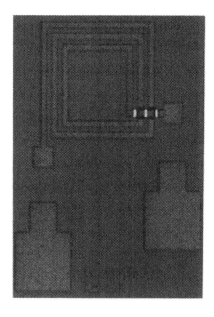

Figure 8.35   *Spiral inductance with air bridges*

8.5.3 Examples of MMIC implementations

Figure 8.36 shows an MMIC preamplifier (Plessey) with associated circuit diagram. Spiral inductances and MESFETs can be clearly identified; the size of the component is only 2.5×1.8 mm^2.

Figure 8.37 shows a photograph of an MMIC wideband amplifier (Harris). The size of the chip is only 1.6×1.8 mm^2. It is worth noting that the noise factor of the MMIC amplifier is higher than with amplifiers with individual components on ceramic substrate. The difference derives largely from the high losses of the GaAs substrate material and the substrate thickness, lower by a factor of four.

As a final example, a transmitter/receiver module for active phased array systems (Figure 8.38) is described. The input signal is amplified via a three-stage amplifier to a pulse output power of several watts, and conducted to the individual transmitter of the antenna. A low-noise preamplifier in the case of the receiver provides for an adequate noise factor and amplification of the module. By means of digitally controlled phase shifters and attenuators (common for transmission and reception), phase and amplitude can be individually adjusted for each module and differently for transmission and reception. There are already layouts for modules of this kind from several companies which are suitable for system implementation. They consist of individual MMIC circuits, such as switches, phase shifters, attenuators, driver amplifiers, and low-noise amplifiers, which in most cases are connected to micros-triplines on ceramic substrate. The control circuit associated with this is integrated in chip form.

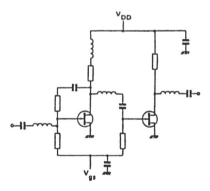

Low-noise MMIC preamplifier circuit diagram.

Figure 8.36 *MMIC amplifier*

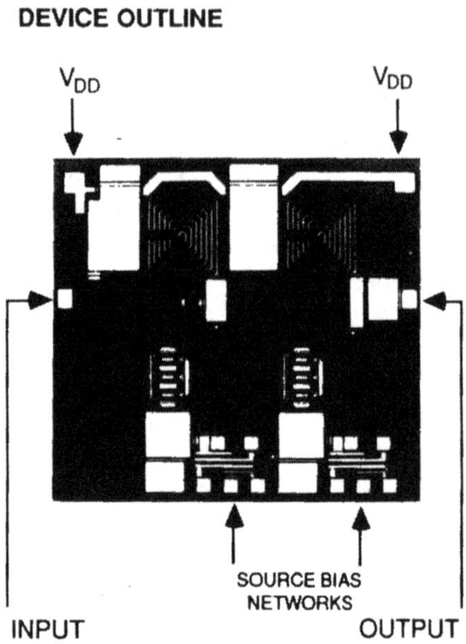

DEVICE OUTLINE

V_{DD} V_{DD}

SOURCE BIAS
NETWORKS

INPUT   OUTPUT

Figure 8.37 *MMIC amplifier*

Figure 8.39 shows a photograph and the block circuit diagram of a four-bit phase shifter with switched bypasses (Plessey). Input attenuation is 1.3 dB per bit, and total phase error is 4°.

MMIC amplifiers are already available today from a number of companies, and the same applies to the SPDT switches. Figure 8.40 shows the layout of a power switch from Plessey. This switch is intended to be able to switch 3 W continuous wave power at 1 dB transmission loss and 20 dB isolation. The amplifier final stage at present still has to be built with a discrete power GaAsFET, since MMICs are not yet available with this output power capacity. The same applies for the transmitter/receiver switch at the output of the module, which in most cases is still implemented as a circulator. These two modules still occupy the largest amount of space on the module.

The T/R module described still consists of, for example, eight individual discrete chips and MMICs. In the final analysis, the trend is still towards creating only two or three multifunction chips in MMIC technology by improving the yield of the circuits, and by greater integration, and by incorporating these chips in a hermetically sealed microwave housing.

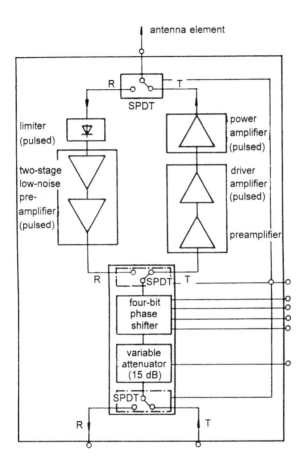

Figure 8.38 *T/R module*

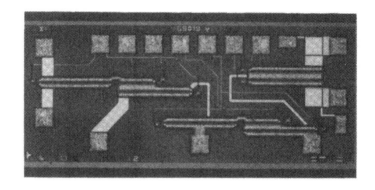

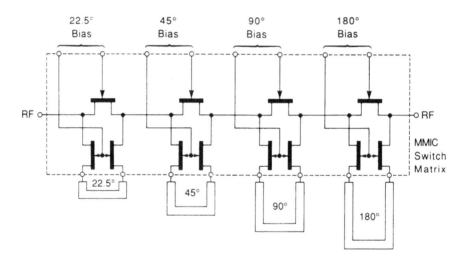

Figure 8.39   *Four-bit phase shifter*

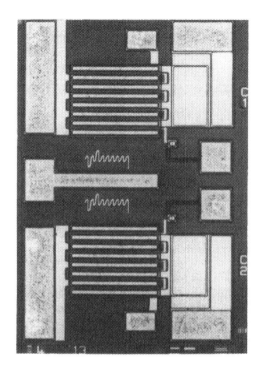

Figure 8.40   *SPDT switch*

Chapter 9

Matching measurement – problems and solutions

G. Lang

HF systems such as telecommunications equipment, radar systems, and navigational systems contain a large number of modules which are connected to one another by lines. The individual modules themselves, and the lines which connect them, are subject to match faults, which cause reflectances and multiple reflectances, and could interfere with the function of the entire system.

The smallest possible faulty matching is desirable for the exact functioning of a system for each module used, and for the connection lines. The production of HF systems would, however, be substantially more expensive, and as a result certain matching errors are permissible in the specifications for the system as a whole and for the individual subsystems.

The modules are monitored with appropriate matching measuring equipment. In this context, it is extremely important to know the error limits of the measuring devices being used to ensure on the one hand that the specifications are maintained, and on the other that no disproportionately high expenditure needs to be incurred during the examination process. Different measuring processes and measuring arrangements also need to be compared with regard to their degree of precision.

9.1 Matching measurement (reflectometry)

To determine the mismatching of a transmission line, the separate measurement of a forward wave and a reflected wave is necessary. This separation is effected by means of measuring systems with directive capability. Figure 9.1 summarises the definition of reflection.

Possible measuring systems with directive capability are directional couplers, VSWR bridges, magic T-circuits, circulators, etc. For swept measurements over a specific frequency range the measuring device should have as little frequency response variation as possible.

In the case of the directional coupler, the not insubstantial frequency response variation is compensated by the use of one directional coupler each for measuring the outgoing and return signals.

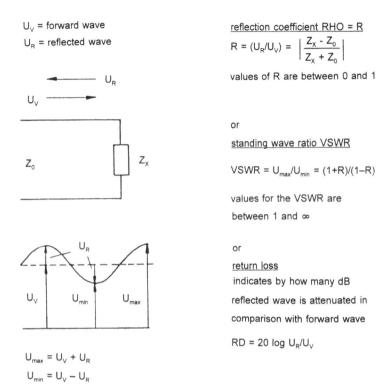

U_V = forward wave

U_R = reflected wave

reflection coefficient RHO = R

$$R = (U_R/U_V) = \left| \frac{Z_X - Z_0}{Z_X + Z_0} \right|$$

values of R are between 0 and 1

or

standing wave ratio VSWR

$$VSWR = U_{max}/U_{min} = (1+R)/(1-R)$$

values for the VSWR are between 1 and ∞

or

return loss

indicates by how many dB reflected wave is attenuated in comparison with forward wave

$$RD = 20 \log U_R/U_V$$

$$U_{max} = U_V + U_R$$

$$U_{min} = U_V - U_R$$

Figure 9.1   *Definition of reflection coefficient* R, *standing wave ratio VSWR, and reflection attenuation* RD

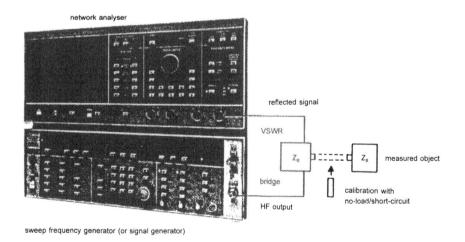

Figure 9.2   *Reflection measurement with scalar network analyser*

The VSWR bridge in general has a relatively greater bandwidth and a smaller frequency response. One disadvantage, however, is the high insertion loss (6 dB for the outgoing wave), which renders these bridges only suitable for lines of less than 500 mW, and makes them incapable of being used in measurement arrangements as operational measuring equipment, as is usual, for example, with radar technology; directional couplers are used for this purpose.

9.2 Reflection ratio

An important factor for a measurement system for matching measurement is the reflection ratio. Figure 9.2 shows a measurement set-up.

In the case of total reflection (open- or short-circuit), at the reflection output of the VSWR bridge a voltage proportional to the outgoing wave is obtained. When terminated with Z_0, a residual voltage is obtained at the reflection output which is caused by asymmetry in the layout, in the balancing process, and by mechanical tolerances. The ratio of the voltage at the reflection output under short-circuit or no-load to the residual voltage E_d at termination with Z_0 is the reflection ratio, quoted in dB. The residual voltage E_d has an effect on the reflection measurement as an error voltage, and adds to the measuring channel signal E_x, as shown in Figure 9.3.

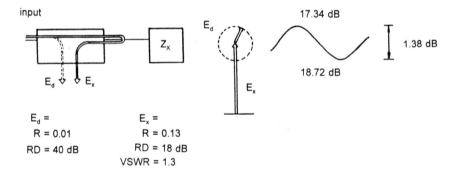

Figure 9.3 *Superimposition of fault signal E_d (40 dB reflection ratio) with measurement signal E_x (VSWR 1.3, RD 18 dB)*

The error signal E_d is superimposed on the measurement signal E_x, and can assume any phase angle. The measured value can lie within the upper limit $E_x + E_d$ and the lower limit $E_x - E_d$. The representation at the scalar network analyser is effected on the dB scale. The difference between the 40 dB reflection ratio of the measuring arrangement and the RD of the measured object of 18 dB is 22 dB, and accordingly E_d is 22 dB smaller than E_x.

The values E_x and E_d for the limit values in dB cannot simply be added, and we accordingly avail ourselves of the VSWR and measuring error table from Figure 9.4.

VSWR $\frac{1+R}{1-R}$	reflection coefficient R	RD = return loss (dB)	dB of small reference value X	maximum $1+X$	minimum $1-X$	ripple $\frac{1+X}{1-X}$
1.134	0.0631	24	24	−0.531	0.566	1.098
1.119	0.0562	25	25	−0.475	0.502	0.977
1.107	0.0501	26	26	−0.434	0.446	0.880
1.096	0.0447	27	27	−0.380	0.397	0.777
1.083	0.0398	28	28	−0.338	0.353	0.691
1.074	0.0355	29	29	−0.303	0.314	0.556
1.065	0.0316	30	30	−0.270	0.279	0.549
1.058	0.0282	31	31	−0.242	0.248	0.490
1.052	0.0251	32	32	−0.215	0.221	0.436
1.046	0.0224	33	33	−0.192	0.197	0.389
1.041	0.0200	34	34	−0.172	0.174	0.347
1.036	0.0178	35	35	−0.153	0.156	0.309
1.032	0.0159	36	36	−0.137	0.138	0.275
1.029	0.0141	37	37	−0.122	0.123	0.245
1.026	0.0126	38	38	−0.109	0.110	0.219
1.023	0.0112	39	39	−0.098	0.098	0.196
1.020	0.0100	40	40	−0.086	0.087	0.173
1.0112	0.0056	45	45	−0.049	0.049	0.097
1.0064	0.0032	50	50	−0.028	0.028	0.056
1.0036	0.0018	55	55	−0.016	0.016	0.031
1.0020	0.0010	60	60	−0.0086	0.0086	0.0172

VSWR $\frac{1+R}{1-R}$	reflection coefficient R	RD = return loss (dB)	dB of small reference value X	maximum $1+X$	minimum $1-X$	ripple $\frac{1+X}{1-X}$
∞	1.00	0	0	−6.00	∞	∞
17.40	0.891	1	1	−5.53	19.28	24.81
8.72	0.794	2	2	−5.08	13.74	18.81
5.85	0.708	3	3	−4.65	10.69	15.38
4.42	0.631	4	4	−4.25	8.66	12.91
3.57	0.562	5	5	−3.87	7.18	11.05
3.01	0.501	6	6	−3.53	6.22	9.75
2.61	0.447	7	7	−3.21	5.14	8.38
2.32	0.398	8	8	−2.91	4.41	7.32
2.10	0.355	9	9	−2.64	3.81	6.45
1.92	0.316	10	10	−2.39	3.30	5.69
1.78	0.282	11	11	−2.16	2.88	5.03
1.67	0.251	12	12	−1.95	2.51	4.46
1.58	0.224	13	13	−1.76	2.20	3.96
1.50	0.200	14	14	−1.58	1.93	3.51
1.43	0.178	15	15	−1.42	1.70	3.12
1.38	0.159	16	16	−1.28	1.50	2.78
1.33	0.141	17	17	−1.15	1.32	2.47
1.29	0.126	18	18	−1.03	1.17	2.20
1.25	0.112	19	19	−0.92	1.03	1.96
1.22	0.100	20	20	−0.83	0.92	1.74
1.196	0.0891	21	21	−0.741	0.811	1.552
1.172	0.0794	22	22	−0.664	0.719	1.382
1.152	0.0708	23	23	−0.594	0.638	1.232

Figure 9.4 *VSWR and measurement error table*

In this Table, the maximum and minimum values of the superimposition of a reference vector 1 are represented with a vector which is smaller by X dB. Because larger signals produce smaller attenuation values, the sign of the maximum values is negative. E_d is 22 dB smaller than E_x; from this it follows that

$$(1 + X) = -\,0.66 \text{ dB and } (1 - X) = 0.72 \text{ dB}$$

However, because E_x is supposed to be 18 dB, the possible limit values for the return loss of the measured object are $18 - 0.66 = 17.34$ dB and $18 + 0.72 = 18.72$ dB.

Figure 9.5 shows the error limits of a VSWR bridge with a reflection ratio of 40 dB as a function of the return loss of the measured object. The lower the return loss of the measured object, the smaller the effect of the rectification ratio on the precision of measurement.

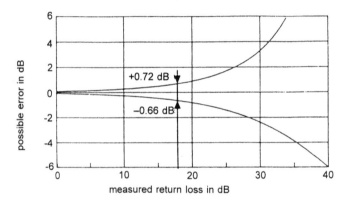

Figure 9.5   *Measurement uncertainty of return loss for reflection ratio of 40 dB*

9.3 Mismatch at measurement output

The second contribution to measurement precision is caused by the mismatch of the measurement output. Although a measurement system may have been adjusted to a high reflection ratio, the internal impedance at the measurement output may deviate from Z_0. In the case of directional couplers, we speak of the VSWR of the main arm (measurement arm). The effects are shown in Figure 9.6.

Calibration is carried out in general with no-load or short-circuit at the measurement output. This causes the outgoing wave to be totally reflected and produces the voltage E_0 at the reflection output. However, the mismatch at the measurement output causes a part of the reflected signal to be reflected once again, which then runs from that point again to the short-circuit or no-load point, where it is totally reflected for a second time, and passes as E_m to the reflection output. E_m can assume any random angle to E_0.

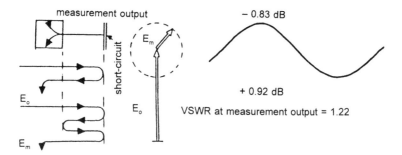

Figure 9.6   *Effect of VSWR at measurement output during calibration*

In the case of a return loss of the measurement output of 20 dB (VSWR = 1.22 or R = 10%), a maximum variation of the output voltage obtains at total reflection of − 0.38 or + 0.92 dB. Calibration with short-circuit or no load alone would therefore already cause a calibration error of maximum 1.75 dB.

The error signal E_m at short-circuit is in antiphase in comparison with no load, which is caused by the doubled reflection. If the calibration data for short-circuit and no load are now determined, the desired value E_0 is obtained as the calibration value. In the scalar network analyser, such mean-value determination is carried out automatically.

Figure 9.7 represents the effect of the same mismatch at the measurement output when measuring an object with a VSWR of 1.3 (return loss = 18 dB or R = 0.126). The measurement signal is attenuated to 18 dB, while the error signal is attenuated to 18 + 20 + 18 = 56 dB. E_m is 38 dB smaller than E_x. The measurement error resulting from this, according to the VSWR table, lies between + 0.11 and − 0.109 dB. The measuring error, caused by the VSWR at the measurement output, increases with the rising return loss of the measured object.

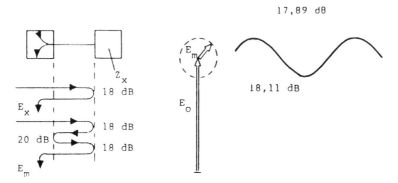

Figure 9.7   *Effect of VSWR at measurement output during measurement*

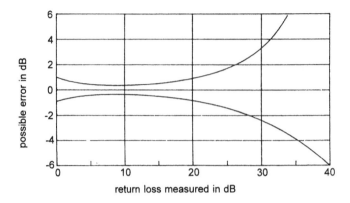

Figure 9.8   *Total uncertainty with VSWR of 1.22 at measurement output and reflection ratio of 40 dB*

The error components caused by the rectification ratio, which is finite, and by the VSWR of the measurement output, can be added or subtracted at will. Figure 9.8 shows the possible total error of a measuring bridge or a directional coupler with a rectification ratio of 40 dB and a VSWR at the measurement output of 1.22 as a function of the return attenuation measured.

9.4 Limits of error in measuring systems

Figure 9.9 shows sets of curves for determining the individual error contributions, depending on the reflection ratio and the mismatch at the measurement output. As a good approximation, the maximum error is obtained by adding the two part errors. The part errors can also be derived from the VSWR table.

For directional couplers, the reflection ratios lie between 26 and 40 dB depending on the bandwidth and frequency. Special directional couplers attain a rectification ratio of up to 50 dB. Conventional bandwidths extend from one octave for standard couplers up to several octaves for special directional couplers. As the bandwidth increases, so the reflection ratio drops. The coupling loss lies between 10 and 20 dB for measurement directional couplers and can vary by up to 6 dB.

In the case of measurement bridges, the reflection ratio lies between 35 and 50 dB, depending on the frequency range. The maximum bandwidth available at present extends from 10 MHz to 40 GHz, at a reflection ratio of 35 dB at 26.5 GHz. The insertion loss is practically frequency independent, thanks to the purely ohmic nature of the layout, being located in theory at 6 dB and in practice between 6 and 7.5 dB. The mismatch at the measurement output is located between 2 and 15 %.

mismatch at

measurement output

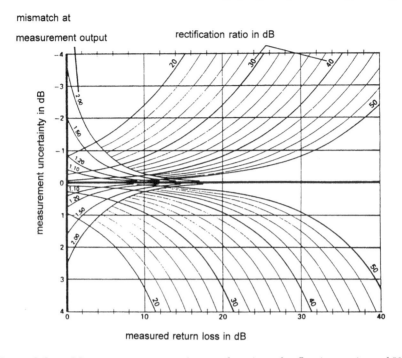

measured return loss in dB

Figure 9.9 *Measurement uncertainty as function of reflection ratio and VSWR at measurement output*

9.5 Errors due to mismatch of signal source

An additional measurement error is engendered by the interaction of the mismatch of the signal source with the mismatch of the measured object. This is demonstrated by way of an example: a signal generator or sweep frequency generator with a mismatch of VSWR = 1.4 is used at the HF output.

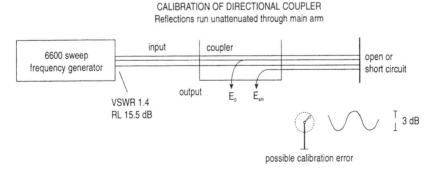

Figure 9.10 *Calibration of directional coupler*

Figure 9.10 shows the signal flow with measurements with a directional coupler. In the case of calibration with short-circuit or no load (corresponding to 0 dB return loss), the forward wave is fully reflected (E_0) and runs unattenuated back to the sweep output, where it is attenuated by 15.5 dB, reflected for the second time, and runs back from there, unattenuated, via the short-circuit as E_{sm} to the reflection output as error voltage. The maximum variation which pertains as a result of this, according to the conversion table, amounts to 3 dB.

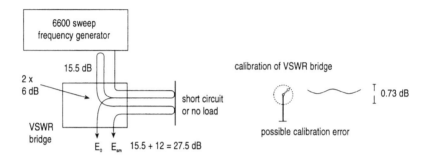

Figure 9.11   *Calibration of VSWR bridge*

Figure 9.11 shows the same calibration with a measuring bridge. The fully reflected signal is attenuated by 6 dB at each pass through the measuring bridge (twice). Together with the return loss from the sweep frequency generator, the error signal E_{sm} is 27.5 dB smaller than E_0. The maximum variation in this case is still 0.73 dB.

An improvement with the directional coupler can be achieved by the intermediate connection of a 6 dB attenuator, although this reduces the measuring dynamics.

9.6 Influence of harmonics

The measurement signal generally also contains harmonics. The influence of the harmonics in the measurement depends on the matching of the measured object and the measuring system with the harmonic. In the most unfavourable scenario, for example with matching measurement for a lowpass filter, the harmonic will be totally reflected. Figure 9.12 shows the effect of the second harmonic on the measurement result. Whether both the limit values can be attained depends on the phase between the fundamental wave and the harmonic. At an interval of 40 dB between measurement signal and harmonic, the maximum variation can amount to 0.17 dB.

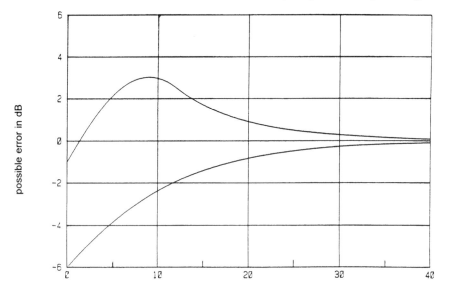

second harmonic in dB below measurement signal

Figure 9.12   *Effect of second harmonic on measurement result*

9.7 Improving measurement precision

When calibrating for the reflection measurement, the effect of the mismatch is greatest at the measurement output, because the calibration is conducted with short-circuit or no-load. Figure 9.13 shows one possibility of eliminating the calibration error for the reflection measurement. The vectors of short-circuit and no-load are in antiphase. The calibration value desired is formed by the means of the values of no-load and short-circuit; for measurements with the scalar network analyser, a mean value determination is carried out, and the frequency response of E_0 is stored. E_0 is subtracted from the measurement result E_x, while at the same time the effect of the frequency response of the measurement set-up is compensated.

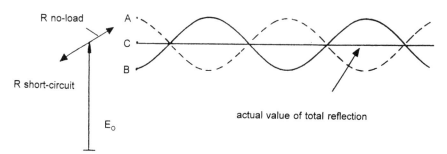

Figure 9.13   *Averaging no-load and short-circuit values*

One important possibility for increasing the measurement precision is, as already discussed, the use of signal generators with good matching to the measurement output. If VSWR measuring bridges are used, this effect is reduced still further. If directional couplers are used, the intermediate connection of an attenuator is recommended.

Figure 9.14 shows the block circuit diagram of the scalar network analyser. The sweepable signal generator used guarantees a VSWR of 1.4 at the measurement output across the entire frequency range from 10 MHz to 18.6 GHz.

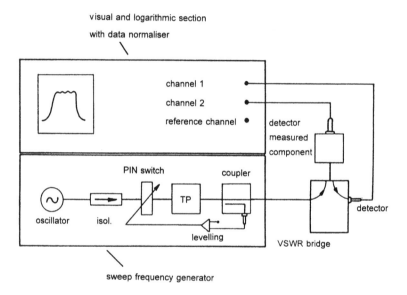

Figure 9.14 *Important components of a measurement circuit*

An isolator is connected after the oscillator, to provide good decoupling between the output and the generator. This is followed by the PIN modulator, and behind that a lowpass filter to reduce the harmonics to less than 40 dB. The excellent VSWR at the measurement output is finally attained by the directional coupler for the levelling, which is located directly at the output. A low-reflection line connects the generator output with the measurement bridge.

9.8 Effect of adapters

A VSWR bridge with a suitable system of plugs is not always available, and in these cases adapters are most often used; this causes the effective reflection ratio and the VSWR at the measurement output of the measurement circuit to be substantially impaired.

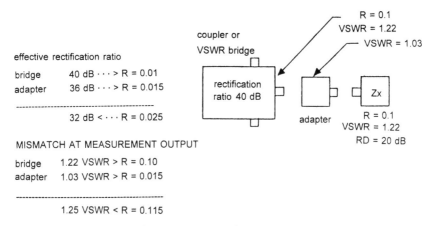

effective rectification ratio

bridge   40 dB · · · > R = 0.01
adapter   36 dB · · · > R = 0.015

----

              32 dB < · · · R = 0.025

MISMATCH AT MEASUREMENT OUTPUT

bridge   1.22 VSWR > R = 0.10
adapter   1.03 VSWR > R = 0.015

              1.25 VSWR < R = 0.115

Figure 9.15   *Effect of adapter on system data*

Figure 9.15 shows, by way of an example, the change in system data of a circuit with 40 dB reflection ratio and VSWR at the measurement output of 1.22. The adapter has a VSWR of 1.03 (R = 0.015; RD = 36 dB). The resultant reflection ratio is then 32 dB and the effective VSWR at the measurement output is 1.25.

When measuring a component with 20 dB return loss (VSWR = 1.22), the following measurement parameters are derived by way of comparison:

measured component	with bridge only	with adapter in addition
20 dB	19.2–20.9 dB	18.1–22.5 dB

9.9 Effect of matching on transmission measurement

It is easy to overlook the fact that matching the measurement circuit also has a not-insubstantial effect on measuring precision during transmission measurement. Figure 9.16 shows a measurement arrangement for measuring the overall attenuation. The individual effects on the measurement result are as follows:

(a) indicates the direction of flow of the measured signal; the measured signal is attenuated by the insertion loss of the measured component.

(b) is the calibration error that occurs owing to reflection between the detector at the measurement output.

(c) arises as a result of the attenuation of the measurement signal in the measured component, return loss at the detector, repeat attenuation at the measured object, return loss at the measurement output, and final attenuation by the measured component.

(d) attenuation by the measured component, reflection at the detector, and reflection at the output of the measured component.

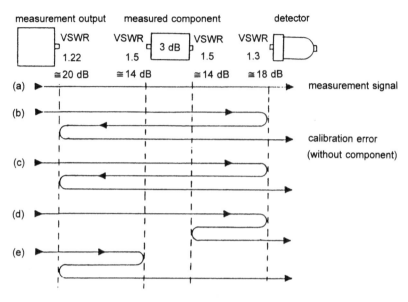

Figure 9.16 *Effects on measuring precision during transmission measurement (throughput attenuation)*

(e) reflection at the input of the measured component, at the measurement output, and attenuation by the measured component.

The intention is to measure a component with attenuation of 3 dB and a VSWR of 1.5 at the input and output. The measurement circuit contains a bridge with a VSWR of 1.22 (return loss = 20 dB) and a detector with a VSWR of 1.3 (return loss = 18 dB). The following signals are to be taken into account:

		dB	R
(i) measured value		3	
(ii) (calibr.) det. + measurement output	$18 + 20$	$= 38$	$\rightarrow$0.013
(iii) 3× attenuation + det. + measurement output	$9 + 18 + 20 = 47$		$\rightarrow$0.005
(iv) attenuation + det. + VSWR	$3 + 18 + 14 = 35$		$\rightarrow$0.018
(v) VSWR + measurement output + attenuation	$14 + 20 + \ \ 3 = 37$		$\rightarrow$0.014

Reflection coefficient, total   26 dB$\leftarrow$0.050

The measured value is 3 dB, and the maximum possible error contribution is 26 dB attenuated. Accordingly, the measured value is 23 dB greater than the error contribution. The measured value, according to the VSWR table can be located between $3 + 0.64 = 3.64$ dB and $3 - 0.59 = 2.41$ dB.

With larger attenuation values, the measurement becomes more precise because the measurement errors become smaller (with the exception of the calibration error).

Small attenuation values should therefore always be measured between measurement points with very good matching.

9.10 Precision measurement with air lines

Measurement processes have been developed with extremely high demands on measurement precision for the measurement of matching and attenuation of wide-band components such as matching resistors, junctions, and attenuators. Measuring the matching in accordance with this process requires a VSWR measuring bridge with four ports. The effective reflection ratio with matching is then 60 dB, with the use of APC 7 plugs in the range between 1 and 18 GHz. If WSMA air lines are used these processes can be used up to 26.5 GHz.

With transmission measurement according to this process, two air lines are needed, which makes it possible to carry out very precise measurements of between 0 and 30 dB attenuation. This process is only practicable with computer evaluation.

9.10.1 Extended measurement of reflection

This method of measurement works in the range between 25 and 55 dB return loss. By using a 20 dB reference at a four-port measurement bridge instead of the terminating resistor Z_0, a large reference voltage is obtained at the output of the bridge, Figure 9.17. The small measurement signal reflected by Z_x is turned in phase by the precision air line, and superimposes itself on the reference signal resulting in ripple. The amplitude is a measure of the reflection of the measured component. It can then be read off from the VSWR table by how many dB the return loss of the measured component is greater than the reference.

If a computer is used for controlling the measurement arrangement, then, with a special measurement program and with the aid of Fourier analysis, the measurement result obtained from the ripple can be printed out directly.

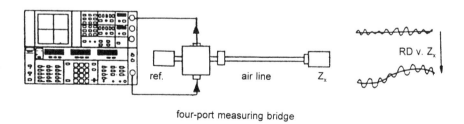

four-port measuring bridge

Figure 9.17 *Arrangement for extended measurement of reflection ratio*

The measurement is carried out as follows:

(1) calibration with short-circuit to the air line (mean value)
(2) connection of Z_x
(3) read off the RD of the reference termination in dB
(4) ripple (peak-to-peak) in dB
(5) conversion of ripple in dB below reference
(6) RD of Z_x = values of (3) and (4) added.

9.10.2 Averaging method for precise determination of reflection

The averaging method (according to Figure 9.18) operates in the range between 0 and 30 dB return loss. In this context, both the calibration error as well as the influence of the reflection ratio of the bridge are eliminated by separating the measured value via the precision air line. By phase rotation of the line, measured value and error are superimposed. This results in a ripple, the mean value of which corresponds to the actual measured value:

— calibration with short-circuit at the air line, mean value
— connection of Z_x; mean value of the ripple is RD from Z_x.

The averaging measurement method can also be automated with the help of Fourier analysis.

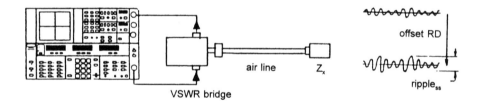

Figure 9.18   *Averaging measurement method*

9.10.3 Precise attenuation measurements between two air lines

As we saw in Section 9.10, the measurement of small attenuation values is corrupted by the VSWR of the connections between which the measurement is made. The smaller the attenuation, the greater this effect becomes. With the help of two precision air lines, however, this error effect can be entirely eliminated. The air lines have an effective return loss of 60 dB. Multiple reflections which occur due to this are of the order of 0.003 dB and can be disregarded. In that case, it is predominantly the precision of the characteristic of the scalar network analyser which is determinant.

Figure 9.19 shows the layout in principle. The evaluation is effected with the same measurement program as the averaging measuring method; in this case too, the measured components must feature a wideband structure. This measuring process is especially well-suited for precision attenuators:

— recording frequency response with two air lines without measured component
— intermediate connection of the measured component and recording of attenuation
— evaluation in the measurement program.

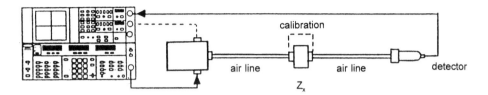

Figure 9.19   *Transmission measurement between two precision air lines*

Chapter 10

Noise and the measurement of noise quantities

N. Krausse

10.1 Introduction

10.1.1 General

Modern receiving systems are frequently required to process very weak signals. Experience has shown it is not possible to make signals of any weakness capable of being processed simply by amplifying them. Instead, the received signal must be of a certain strength, because the interference power created in the receiving system (the noise) overlays the received signal, and may even smother it.

Sensitivity and noise factor are the conventional parameters which describe the capability of a system to process weak signals. The noise factor is particularly important, since it characterises not only the system as a whole but also its modules (such as HF preamplifiers, mixers, or IF amplifiers). By making a careful selection of the noise factor and the amplification of individual stages, the designer can optimise the noise factor and the dynamics of the system as a whole. With a known noise factor, the system sensitivity can be easily estimated from the system bandwidth.

There are three main possibilities for obtaining a reception signal at the output of the receiver system which has the minimum signal-to-noise ratio desired:

– a sufficiently sensitive receiver, and /or
– a sufficiently high transmitting power, and/or
– an antenna with a sufficiently large aperture.

Of these three, the first possibility is preferred in general, not least because it is also the most economical. Accordingly, one needs to determine the noise characteristics of a network unambiguously, precisely, quickly, and in a way that can be repeated. The noise characteristics of oscillators are not considered: these are characterised not by a 'noise factor', but by the 'single sideband phase noise in a bandwidth of 1 Hz at a deviation of x Hz from the carrier'.

10.1.2 Signal-to-noise ratios

In addition to the effective output power that relates to a transmitted message, interference power always occurs. Subject to the precondition that the interference signal exists in the form of noise, the ratio of the effective power to the interference power at the same location is the signal-to-noise ration S/N, observed at that location. For example, the following values apply to the S/N at a loudspeaker, or at a television tube:

	S/N (dB)
lower limits for speech comprehension	10
adequate music reproduction quality	30
adequate television picture quality	40

In the case of digital communication systems, the reliability of the transmission is specified by the error probability P(e). P(e) indicates whether a received bit is erroneous. Figure 10.1 shows P(e) as a function of the carrier/signal-to-noise ratio for various types of digital modulation.

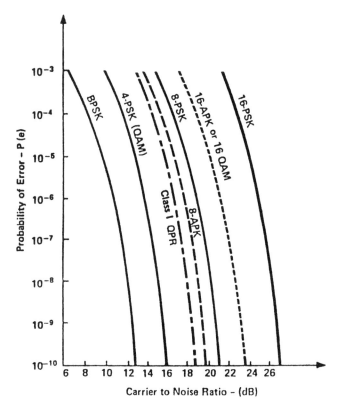

Figure 10.1 *Probability of error as a function of carrier-to-noise ratio for various types of digital modulation*

10.2 Noise sources

10.2.1 General

The statistical fluctuations of voltages and currents (noise) that are superimposed on useful signals have two different main causes:

— external noise sources, the power of which is absorbed by the receiver antenna simultaneously with the reception signal. Such external noise sources are, for example, atmospheric interference, the thermal noise from the earth or from a number of heavenly bodies or star systems. These external noise sources do not fall within the scope of this Section.

— internal noise sources, which are located within the receiver system. In this case, we encounter the thermal noise of the resistors and the various noise phenomena in the semiconductors (shot noise, partition noise, etc.).

The first point to consider in this Section is thermal noise, the most important of these phenomena.

10.2.2 Thermal noise

If the temperature of an equivalent resistance R (hereafter referred to simply as resistance) is above the absolute zero point, in other words 0 K, then noise voltage will occur between the connections of the resistance. The causes of this are the turbulent motions which the atoms of the conductor material perform under the effect of heat. The free electrons of the resistance, which as a total entity represent the electron gas that is present in the resistance, are knocked by the turbulent motion of the atoms, which results in irregular zig-zag movements of the electrons. This in turn causes constantly changing loads on the two resistance connections, and this produces the noise voltage.

The relationships which are important with regard to the thermal noise are derived from Planck's radiation formula. In 1900, Planck produced a famous formula for radiation, which described the spectral distribution of power from the aperture of a black body:

$$\frac{dS}{df} = \frac{2}{\lambda^2} \cdot \frac{h \cdot f}{e^{\frac{hf}{kT}} - 1} \tag{10.1}$$

where

df	= small frequency step (Hz)
f	= frequency (Hz)
dS	= radiation density (W/m^2) in the step df
h	= Planck's constant $6.62 \cdot 10^{-34}$ (Ws^2)
k	= Boltzmann's constant $1.38 \cdot 10^{-23}$ (Ws/K)
λ	= wavelength (m)
T	= temperature (K)

Figure 10.2 shows a radiating surface da. At a distance r is a receiving surface da′ = da. This is perpendicular to the direction of radiation; let the dimensions of transmitting and receiving surfaces be small in relation to the distance interval r.

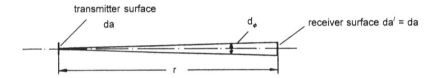

Figure 10.2   *Transmitter and receiver surfaces*

The radiated power dWs impinging on the receiver surface is proportional both to the solid angle dφ = da′/r^2 as well as to the size of the transmitter surface. Introduce a proportionality factor S, and write

$$dWs = Sd \cdot \varphi \cdot da \tag{10.2}$$

S is referred to as the radiation density of the transmitter. Now make use of an antenna from which the received power is conducted via a rectangular hollow conductor and a hollow conductor filter to a matched terminating resistor (Figure 10.3). Let the hollow conductor filter be ideal, with zero attenuation in the passband and attenuation outside the passband.

Let the radiation density of the transmitter be dS within df. The rectangular hollow conductor ensures that, of the randomly polarised transmitter power in the interval df, for example, only the vertically polarised component passes to the filter and onto the terminating resistor (one degree of freedom).

$$dWe = (1/2) dWs$$

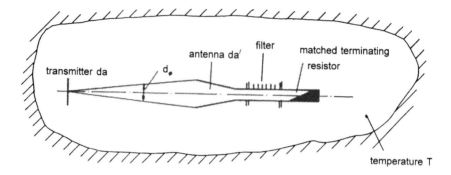

Figure 10.3   *Radiating surface and antenna with terminating resistor*

In other words, the power which reaches the absorber is

$$dWe = \frac{1}{2} \, ds \cdot d\varphi \cdot da \qquad (10.3)$$

Let the vertically polarised component of the radiation and the radiation outside df in this case be entirely reflected to the transmitter. Assume that the system, consisting of transmitter and receiver, is at a constant temperature T; in other words, that no heat can be dissipated to the outside. This is indicated by the envelope in Figure 10.3.

According to the second law of thermodynamics, neither transmitter nor receiver can gain energy at the expense of the counterpart, and both parts of the system must therefore feature the same temperature T. On the other hand, the transmitter constantly radiates the power (1/2) dWs to the receiver. It follows that the terminating resistor must radiate the same power (1/2) dWs back again to the transmitter.

The receiving antenna must have a directional characteristic of such a nature that it radiates the power derived from the absorber solely onto the transmitter antenna; in other words, in such a way that nothing gets lost. For antenna gain, the following applies on the one hand:

$$G = \frac{4\pi \cdot da'}{\lambda^2}$$

but also

$$G = \frac{4\pi}{d\varphi}$$

and therefore

$$dj = \frac{\lambda^2}{da'} = \frac{\lambda^2}{da} \qquad (10.4)$$

Substituting eqn. 10.4 into eqn. 10.3 gives

$$dW_e = \frac{1}{2} dS \cdot \lambda^2 \qquad (10.5)$$

From eqns. 10.1 and 10.5 it follows that:

$$dW_e = \frac{hf}{e^{\frac{hf}{kT}} - 1} \cdot df \qquad (10.6a)$$

Equation 10.6 tells us what power a resistor gives off at the temperature T. Worth noting is the fact that the size of the resistor does not play a part.

The only condition for the output of power dW_e is the matching. Matching means that the source resistance, the wave impedance of the transmission line, and, finally, the terminating resistor must all have the same value.

Examine the expression hf/kT. The values were

$$h = 6.62 \cdot 10^{-34} \ (Ws^2)$$
$$k = 1.38 \cdot 10^{-23} \ (Ws/K)$$

If $f = 100$ GHz (10^{11} Hz) and $T = 290$ K, then

$$\frac{hf}{kT} = \frac{6.62 \cdot 10^{-34} \cdot 10^{11}}{1.38 \cdot 10^{-23} \cdot 290} \ll 1$$

e^x for small x can be substituted by $1 + x$. One can see that up to about 100 GHz eqn. 10.6 can be replaced by

$$dW_e = kT \ df \qquad (10.6b)$$

Table 10.1 illustrates eqn. 10.6a.

Table 10.1   *Evaluation of eqn. 10.6a*

f(Hz)	dW_e (dBm) df = 1 (Hz) T_0 = 290 (K)
10^1	-174.0
10^2	-174.0
10^3	-174.0
10^4	-174.0
10^5	-174.0
10^6	-174.0
10^7	-174.0
10^8	-174.0
10^9	-174.0
10^{10}	-174.0
10^{11}	-174.0
10^{12}	-174.3
10^{13}	-178.0
10^{14}	-233.4
10^{15}	-867.5

The noise power P_n was

$$dW_e = P_n = kT\,df \tag{10.7}$$

The energy kT for T_0 = 290 K was

$$kT_0 = 1.38 \cdot 10^{-23} \cdot 290 = 4 \cdot 10^{-21} \text{ (W/Hz)} \tag{10.8a}$$
$$kT_0 = -174 \text{ dBm (related to 1 Hz bandwidth)} \tag{10.8b}$$

A mean noise voltage square (Figure 10.4) corresponds to this noise power:

$$U^2 = kT\,R\,df \tag{10.9}$$

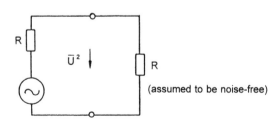

Figure 10.4   *Noise voltage*

Observing (amplified) resistance noise on an oscilloscope, one can see that higher noise voltage amplitudes (noise peaks) are very rare. In fact, the probability w of a particular voltage value U being exceeded is described by

$$w = 1 - \Phi\left(\frac{U}{\sqrt{\overline{U^2}} \cdot \sqrt{2}}\right) \qquad (10.10)$$

Φ designates the Gauss error integral. As seen from Table 10.2 and Figure 10.5, the probability of greater values of U diminishes extraordinarily quickly. Even values larger than 3.5 times the squared mean value occur on average only once in 2000 values.

Table 10.2   *Evaluation of eqn. 10.10*

$\dfrac{U}{\sqrt{\overline{U^2}}}$	1	1.5	2	2.5	3	3.5	4
w	31.7%	13.4%	4.6%	1.24%	0.27%	0.05%	0.006%

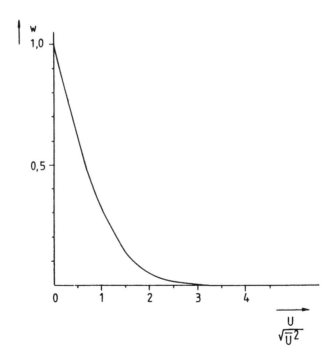

Figure 10.5   *Graphic representation of Table 10.2*

10.2.3 Shot noise

10.2.3.1 Shot current
It is well known that electric current is not a continuous phenomenon, but consists of discrete electrons of charge

$$e = 1.6 \cdot 10^{-19} \text{ (Cb)}$$

In addition, the electron flow is not absolutely uniform, but is irregular in its magnitude. If a direct current I_0 is flowing, this is a mean value, and tells us nothing about the variation of the momentary value and the frequencies associated with it.

It can be shown that the variations of the mean square of the current are distributed uniformly in the frequency:

$$I_s^2 = 2 \, e \, I_0 \, \Delta f \qquad\qquad (10.11)$$

$$
\begin{aligned}
e \quad &= 1.6 \cdot 10^{-19} \text{ (Cb)} \\
I_0 \quad &= \text{direct current (A)} \\
f \quad &= \text{bandwidth (Hz)}
\end{aligned}
$$

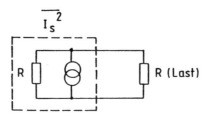

Figure 10.6   *Shot noise source*

Equation 10.11 applies to frequencies with a period substantially greater than the run time of the electrons through the component in question. The shot noise power available (Figure 10.6) is calculated from eqn. 10.11:

$$P = \frac{1}{2} e \, I_0 \, \Delta f \, R \qquad\qquad (10.12)$$

As an example, $I_0 = 0.1$ A and $R = 100 \, \Omega$. The energy accordingly becomes $W = 0.8 \times 10^{-18}$ (W/Hz).

10.2.3.2 Other phenomena
There are other phenomena which occur in active components (particularly semi-conductors) which can be treated in a similar statistical manner to shot noise, thanks to their quantum nature. For example,

— the generation and recombination of electron–hole pairs in semiconductors

— the partition of the emitter flow between base and collector in bipolar transistors ('current partition noise').

10.2.4 Noise from IMPATT diodes

Impact avalanche transit time (IMPATT) diodes represent a particularly important source of noise, since diodes of this type are well suited in particular for use as a reference source for noise factor measurements.

With avalanche diodes, a charge carrier acquires so much energy owing to the imposed DC voltage that, when it collides with the crystal lattice, one or more electron–hole pairs are produced; these pairs then produce further charge carriers, and so on (the avalanche effect). The multiplication factor in the production of free load carriers can be distributed at will, as well as being quantised. The noise power of an avalanche diode in the usable range is approximately inversely proportional to the current, and, to a lesser degree, frequency dependent.

10.3 Noise factor

10.3.1 General

The value most used for characterisation of the noise properties of a linear two-port network is the noise factor. Harold Friis defined the noise factor in 1944 as follows:

$$\text{noise factor F} = \frac{\text{signal–to–noise ratio at the input}}{\text{signal–to–noise ratio at the output}} \qquad (10.13)$$

The noise factor is therefore a measure for the reduction or impairment of the signal-to-noise ratio when processing the signal in the network in question. An amplifier, for example, does not only amplify the signal and noise pertaining at its input, but adds a specific noise component to them. The example in Figure 10.7a shows the spectral representation of the ratios at the input of an amplifier. The signal is 40 dB higher than the noise floor.

Figure 10.7b shows the ratios at the amplifier output. The signal is raised by the amplification of the two-port network (20 dB). However, the signal is still only 30 dB above the noise floor. According to the definition (eqn. 10.13), one can say that the amplifier has a noise factor of 10 dB.

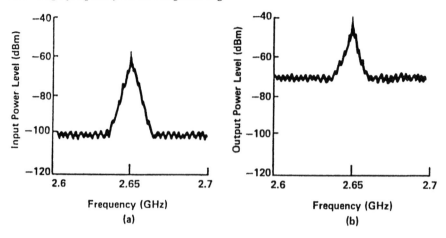

Figure 10.7   *Typical signal and noise ratios at input and output of amplifier*

10.3.2 Definition of noise factor F

The principal definition is now considered in greater detail. According to eqn. 10.13,

$$F = \frac{S_i/N_i}{S_o/N_o} = \frac{N_o}{N_i S_o/S_i}$$

S_i = signal input power
N_i = noise input power
S_o = signal output power
N_o = noise output power

The available gain (Section 10.6) of a two-port network is $G_a = S_o/S_i$ and therefore

$$F = \frac{N_o}{G_a N_i} \tag{10.14}$$

The noise output power N_o is composed of the amplified input noise $G_a N_i$ and the noise N_a produced in the two-port network itself:

$$N_o = G_a N_i + N_a \tag{10.15}$$

Accordingly, eqn. 10.14 becomes

$$F = \frac{G_a N_i + N_a}{G_a N_i} \tag{10.16}$$

The noise input power N_i is normally the thermal noise of the source resistance, eqn. 10.7:

$$N_i = kTB \tag{10.17}$$

B = noise bandwidth (Hz). One can see that, because of the connection of eqn. 10.17, the noise factor (eqn. 10.16) is a function of the source temperature T.

To guarantee comparable ratios, a temperature of $T_0 = 290$ K has been adopted by the IRE as the standard temperature for the definition of noise factor:

$$F = \frac{N_a + kT_0BG_a}{kT_0BG_a} \tag{10.18}$$

The noise factor of a two-port network is, in general, a function of frequency, but, subject to the precondition of a narrow measuring bandwidth (Figure 10.8), is largely independent of the bandwidth. The two noise power values N_i and N_a of eqn. 10.15 are both proportional to the bandwidth, with the result that in the final analysis the noise factor is independent of the bandwidth.

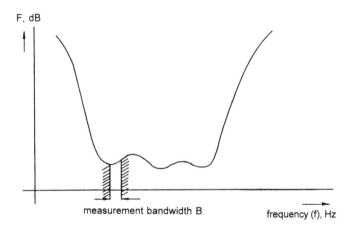

Figure 10.8   *System and measurement bandwidth*

10.3.3 Effective input noise temperature T_e

A noisy two-port network can be visualised as being replaced by a noise-free arrangement. In this case, an internal equivalent noise power source R_s is then assumed at the input of the two-port network, with the temperature T_e (Figure 10.9), the noise power of which supplies the additional noise power N_a after amplification by G_a.

T_e denotes the effective input noise temperature of a two-port network. T_e, in addition to the noise factor F, represents the second possible method of describing the noise characteristics of a two-port network. The advantage is that no temperature reference T_o is required. By definition, the following applies to the noise created in the two-port network:

$$N_a = G_a k T_e B \qquad (10.19)$$

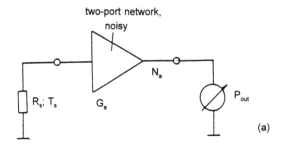

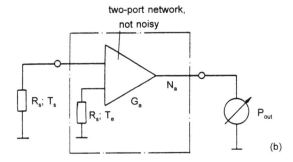

Figure 10.9 *Effective noise temperature of a two-port network*

Inserted in eqn. 10.18 it follows that

$$F = 1 + \frac{T_e}{T_o} \qquad (10.20)$$

as the relation between T_e and F. Satellite and space travel applications require reception systems with particularly small noise factors. In these cases, it is conventional to indicate the noise temperature T_e instead of the noise factor F.

Example: Let a GaAs FET amplifier cooled by liquid helium have the noise temperature $T_e = 20$ K. It follows that

$$F = 10\log\left(1 + \frac{20}{290}\right) = 0.29 \text{ (dB)}$$

10.3.4 System noise temperature, system noise factor

In a realistic receiver system, the antenna represents a noisy source resistance (Figure 10.10). The noise temperature for terrestrial systems amounts to about 290 K (temperature of the ground, building structures, etc.), and for space flight applications approximately 4 K (background temperature of outer space). According to Figure 10.10, the system noise temperature is simply:

$$T_{sys} = T_a + T_e \tag{10.21}$$

Occasionally, the term system noise factor is used:

$$F_{sys} = \frac{T_{sys}}{T_o} = \frac{T_a + T_e}{T_o} \tag{10.22}$$

It should be borne in mind that the system noise factor under ideal conditions $(T_a = 0 \text{ K}, T_e = 0 \text{ K})$ becomes $0 \ (= -\infty \text{ dB})$.

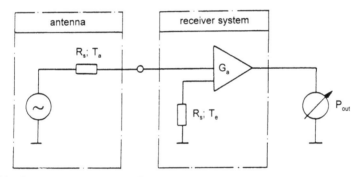

Figure 10.10 *Receiver system with antenna*

10.3.5 Ladder network of noisy two-port networks

A receiver system normally consists of several stages with different amplification values and noise factors (Figure 10.11). The total noise factor of the arrangement is now of interest.

The power available at the output of the first stage is

$$P_{a1} = k \cdot T_s \cdot B \cdot G_{a1} + k \cdot T_{e1} \cdot B \cdot G_{a1} \tag{10.23}$$

The power available at the output of the second stage is

$$P_{a2} = P_{a1} \cdot G_{a2} + k \cdot T_{e2} \cdot B \cdot G_{a2} \tag{10.24}$$

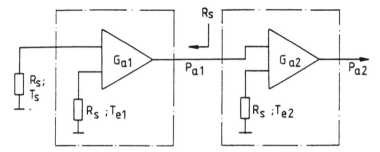

Figure 10.11   *Cascade arrangement*

and with eqn. 10.23

$$P_{a2} = k \cdot T_S \cdot B \cdot G_{a1} \cdot G_{a2} + k \cdot T_{e1} \cdot B \cdot G_{a1} \cdot G_{a2} + k \cdot T_{e2} \cdot B \cdot G_{a2}$$
(10.25)

Regarding the cascade arrangement from Figure 10.11 as a single stage with amplification G_{a12} and effective noise temperature T_{e12}, then

$$P_{a2} = k \cdot T_S \cdot B \cdot G_{a12} + k \cdot T_{e12} \cdot B \cdot G_{a12}$$
(10.26)

In view of the fact that $G_{a12} = G_{a1}G_{a2}$, there follows in the final analysis, by equating eqns. 10.25 and 10.26,

$$T_{e12} = T_{e1} + \frac{T_{e2}}{G_{a1}}$$
(10.27)

With the relationship of eqn. 10.20

$$F = 1 + \frac{T_e}{T_o}$$

the total noise factor is finally

$$F_{12} = F_1 + \frac{F_2 - 1}{G_{a1}} \; ...$$
(10.28)

(Friis noise formula)

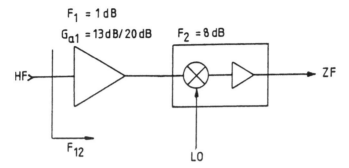

Figure 10.12 *Example of application of Friis noise formula*

Example: Consider a ladder network consisting of a low-noise HF preamplifier and a mixer/IF preamplifier (Figure 10.12). Let the initial values be

$$F_1 = 1.0 \text{ dB} = 1.259$$
$$F_2 = 8.0 \text{ dB} = 6.310$$
$$G_{a1} = 13.0 \text{ dB} = 20.0$$

From this, calculate with eqn. 10.28

$$F_{12} = 1.259 + \frac{6.31 - 1}{20} = 1.524 = 1.82 \text{ dB}$$

One can identify a drastic deterioration in the total noise factor in relation to the noise factor of the HF preamplifier. If the amplification G_{a1} is increased to 20 dB (= 100), it follows that

$$F_{12} = 1.259 + \frac{6.31 - 1}{100} = 1.312 = 1.18 \text{ dB}$$

10.3.6 Typical noise factors

Figure 10.13 shows typical noise temperatures/noise factors of some transistors in comparison with cooled arrangements (using liquid helium) and in comparison with the noise temperature of the sky at zenith.

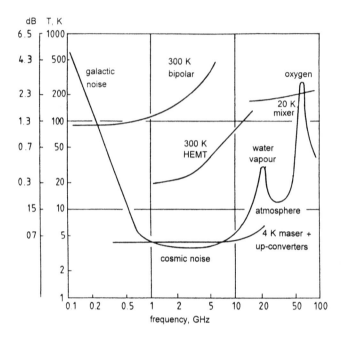

Figure 10.13 *Typical noise factors*

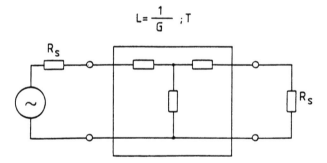

Figure 10.14 *Matched attenuator*

10.4 Noise from components

10.4.1 Matched attenuators

Figure 10.14 shows a matched attenuator with loss $L = 1/G$. The attenuator is at temperature T. An attenuator of this kind can, for example, be a conducting cable between antenna and receiver system. It can be seen that the following applies to the effective input noise temperature of such an attenuator:

$$T_e = T (L - 1) \tag{10.29}$$

and with eqn. 10.20 for the noise factor

$$F = 1 + \frac{T}{T_0} (L - 1) \tag{10.30}$$

For the standard case $T = T_0$, it follows that

$$F = L = \text{attenuator loss} \tag{10.31}$$

Line losses accordingly have a direct effect on the system noise factor. This means that in cases with especially high sensitivity requirements the receiver is arranged directly behind the antenna.

10.4.2 Mixer diodes

In many receiver systems, with the aid of the mixer, the (generally high) signal frequency f_s is mixed with the help of the oscillator frequency f_{LO} into the position of a (generally low) intermediate frequency f_{IF}. In principle, the mixer is in a position to process two different signal frequencies $f_{LO} + f_{IF}$ and $f_{LO} - f_{IF}$ (double sideband operation), Figures 10.15a, b. One of the two frequencies is the signal frequency f_s which is required, and the other is the (generally undesirable) image frequency f_{sp}.

The signal processing and measurement arrangements, and the mathematical relationships dealt with within the scope of this Section are, however, subject to the usual single-sideband reception conditions; i.e. reception on the image frequency is prevented by appropriate measures (e.g. image selection filters; Figures 10.16a, b). As far as noise factor considerations are concerned, the mixer constitutes an imperfect component. Let the mixing loss be L. The noise factor of a mixer, however, is not identical to its mixing loss. Rather, the noise factor is

$$F = L \, t_m \tag{10.32}$$

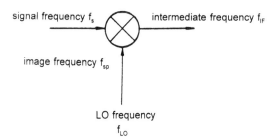

Figure 10.15a *Receiver mixer without image selection*

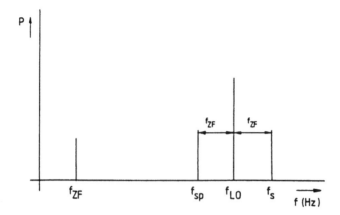

Figure 10.15b *Frequency components for mixer*

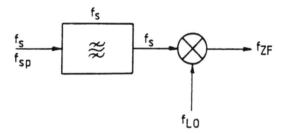

Figure 10.16a *Receiver mixer with image selection*

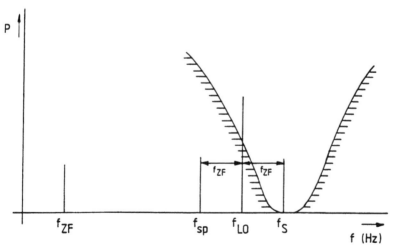

Figure 10.16b *Spectrum of receiver mixer with image selection*

t_m denotes mixer noise ratio. For the situation that often arises in which the image frequency requires matching:

$$t_m = \frac{2}{L}\left(\frac{T_d}{290}\left(\frac{L}{2} - 1 \right) + 1 \right)$$ (10.33)

T_d is the noise temperature of the diode. From eqns. 10.32 and 10.33 there follows:

$$F = \frac{T_d}{290}(L - 2) + 2$$ (10.34)

With $T_d = 290$ K, naturally $F = L$. In general, however, T_d amounts to about 350 K.

With a typical mixing loss of $L = 6$ dB and $T_d = 350$ K, according to eqn. 10.34, $F = 4.41 = 6.4$ dB. For practical applications, then, the approximation $F = L$ is sufficient.

10.4.3 Transistors

10.4.3.1 General
The intention in Sections 10.4.3.2 and 10.4.3.3 is to demonstrate briefly which parameters of a transistor have an effect on their noise behaviour. In practice the transistor is characterised in a manner appropriate to the purpose by its noise parameters F_{min}, $G_0 + jB_0$ and R_n (see Chapter 6, Section 6.3).

10.4.3.2 Bipolar transistors
To estimate the minimum noise factor F_{min} of a bipolar transistor, the following applies for sufficiently high frequencies:

$$F_{min} \approx 1 + h\left(1 + \sqrt{1 + \frac{2}{h}} \right)$$ (10.35)

with

$$h = \frac{q \cdot I_c \cdot r_b}{k \cdot T}\left(\frac{f}{f_T} \right)^2$$

or

$$h = 0.04 \times I_c \times r_b \left(\frac{f}{f_T} \right)^2$$ (10.36)

I_C = collector current (mA)
r_b = basic resistance (Ω)
f_T = cut-off frequency ($|h_{21}| = 1$)
T = 290 K

The noise factor F_{min} cannot of course simply be made as small as desired as a result of this, because I_C is constantly being reduced. Rather, as I_C decreases, f_T decreases also, with the result that the noise factor reaches a minimum at a specific I_C.

Example for calculation of F_{min}: Let the transistor chip HP HXTR-7011 be given as

V_{CE} = 10 V
I_C = 10 mA
f_T = 7.5 GHz
r_b = 3 Ω

Accordingly, at 2 GHz with eqn. 10.36, $h = 0.085$, and from eqn. 10.35, $F = 1.8$ dB. The HP data sheet gives 1.7 dB.

10.4.3.3 Field effect transistors

To estimate the minimum noise factor F_{min} of a MESFET, the following applies for room temperature:

$$F_{min} \approx 1 + k_3 \cdot L \cdot f\sqrt{g_{mo}} \, (R_S + R_G) \tag{10.37}$$

with

k_3 = 0.3
L = gate length (μm)
f = frequency (GHz)
g_{mo} = transconductance (S)
R_S = internal source resistance (Ω), Figure 10.17
R_G = internal gate resistance (Ω), Figure 10.17

As an example, take the MESFET NE71000 (Figure 10.17), gate length 0.3 (μm). The agreement of the noise factors in Table 10.3 according to the data sheet and the approximation according to eqn. 10.37 is good.

Table 10.3 *Comparison of noise figures for MESFET*

f (GHz)	F (dB) datasheet	F (dB) as per eqn. 4.9
4	0.6	0.6
12	1.6	1.6
18	2.1	2.2

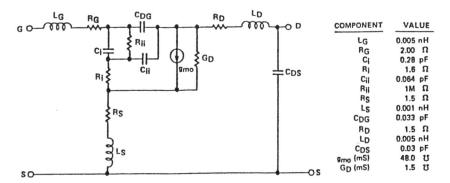

COMPONENT	VALUE
L_G	0.005 nH
R_G	2.00 Ω
C_i	0.28 pF
R_i	1.6 Ω
C_{ii}	0.064 pF
R_{ii}	1M Ω
R_S	1.5 Ω
L_S	0.001 nH
C_{DG}	0.033 pF
R_D	1.5 Ω
L_D	0.005 nH
C_{DS}	0.03 pF
g_{mo} (mS)	48.0 $\mho$
G_D (mS)	1.5 $\mho$

Figure 10.17 *Equivalent circuit diagram of GaAs MESFET NE71000*

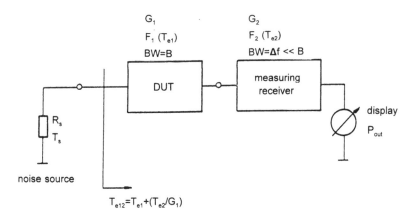

Figure 10.18 *Basic arrangement for noise factor measurement*

10.5 Measuring noise factor

10.5.1 Level plan

Figure 10.18 shows the basic arrangement for a noise measurement. A calibrated noise source with the internal resistance R_S and the temperature T_S (in most cases capable of being switched between two values) serves as reference.

Let the measured object (DUT = device under test) have the gain G_1, the noise factor F_1 (or the noise temperature T_{e1}) and the bandwidth B (Hz). Because the noise power of the noise source R_S is small, and the DUT in most cases features only relatively low loss, a measurement receiver is in most cases required for a reliable display of the output power P_{out}. Let this measurement receiver have amplification G_2, the noise factor F_2, and the bandwidth f (Hz) (small in relation to the bandwidth B of the DUT). The measurement receiver can operate with or without frequency conversion. To estimate the amplification required G_2 for the measurement receiver, consider the noise temperature at the input of the DUT $T_{tot} = T_S + T_{e12}$, with

$$T_{e12} = T_{e1} + \frac{T_{e2}}{G_1} \qquad\qquad \text{(as 10.17)}$$

Then

$$P_{out} = k\, T_{tot}\, \Delta f\, G_1\, G_2 \qquad\qquad \text{(10.38)}$$

An example may help explain eqn. 10.38. Let these values be given:

F_1 = 3 dB = 2
F_2 = 10.4 dB = 11
G_1 = 20 dB = 100
Δf = 1 MHz
T_S = 34 T_0 (see Section 10.5.2.1)

We seek the amplification G_2 of the measurement receiver, for $P_{out} = 1$ (mW) = 0 dBm. With eqns. 10.20 and 10.17, $T_{e12} = 1.1 \cdot T_0$. Therefore $T_{tot} = 35.1 \cdot T_0$. For P_{out} with $\Delta f = 1$ (MHz), we can state

$$P_{out} = 0 \text{ (dBm)} = -174 + 60 + 10 \log 35.1 + 20 + G_2$$

and from this $G_2 = 78.5$ dB.

10.5.2 Measuring devices for measuring noise factors

10.5.2.1 Noise sources
A calibrated noise source is needed as a reference to measure the noise factor. Arrangements can be considered as measurement noise sources that will allow the noise temperature to be switched over to two precisely defined values. The following arrangements have been or are still used:

- thermionic diodes
- gaseous discharge tubes
- avalanche diodes
- heated or cooled resistors ('hot–cold standard').

Up to the end of the 1950s, popular noise sources were of such a kind that their noise behaviour could be derived in theory, because noise source calibration processes were not yet available.

Thermionic diodes make use of the shot current and the noise power associated with this. Because of the restricted frequency ranges (up to about 100 MHz) and the high DC voltage required for operation, the thermionic diodes are no longer used.

Gaseous discharge tubes (with cold cathodes) were a widespread noise source, from UHF through to millimetric wavelengths. In this context, the tube is arranged at an angle in a hollow conductor. A disadvantage with these tubes is that they require an ignition voltage of several thousand volts and an operational voltage in the range 100–300 V.

The avalanche diode (see Section 10.2.4) has completely displaced the gaseous discharge tube. Manufacturers of measuring equipment offer noise diodes in which one single model can cover the entire frequency range from 10 MHz to 26.5 GHz.

Heated or cooled resistors are used as calibration standards. The 'hot–cold standards' available on the market use two terminators, one of which is cooled by liquid nitrogen to 77 K, and the other is thermostatically regulated in an oven to 373 K.

10.5.2.2 Measurement receivers
The precision of noise factor measurement is determined not only by the properties of the noise source, but also by the uncertainty of measurement of the measurement receiver. The measurement receiver must possess the following characteristics:

- correct signal frequency
- adequate amplification
- adequate dynamics
- maximum linearity
- high stability of amplification

A desirable property is a low inherent noise factor. Figure 10.19 shows the block circuit diagram of a typical measurement receiver. The heart of the arrangement is the fixed-frequency receiver for an IF of, for example, 30 MHz, such as a number of manufacturers supply. This receiver contains a precision attenuator ATT. With the ATT, the increase in noise power after 'ignition' is balanced out with the noise source, so that the rectifier CR always operates at the same working point.

Signal frequencies $f_S >$ IF are mixed into the level of the IF. To do this, one makes use of a low-noise HF preamplifier, an image selection filter to guarantee the sideband operation required from Section 10.4.2, and the up converter.

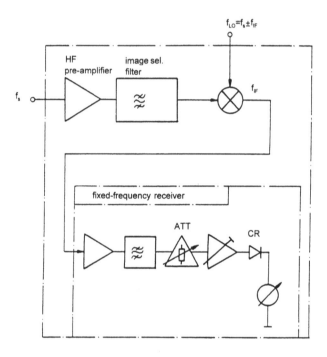

Figure 10.19 *Measurement receiver for noise factor measurement*

10.5.3 Signal generator method

The signal generator method is nowadays used only for measuring larger noise factors (> 30 dB). Figure 10.20 shows the measuring arrangement. In the first instance, only the noise power N_i of the source resistance R_S (temperature T_0) is conducted to the DUT.

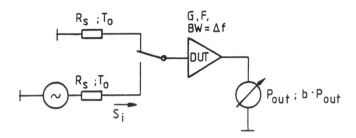

Figure 10.20 *Signal generator method*

The following applies at the output

$$P_{out} = N_o = G\,N_i + N_a \qquad (10.39)$$

The signal power S_i is now also imposed at the input. As a result of this, the output power rises by the factor b (measurement with the precision attenuator ATT, Figure 10.19):

$$b\,P_{out} = G\,S_i + N_o \qquad (10.40)$$

and because of eqn. 10.39

$$N_o = \frac{G\,S_i}{b-1} \qquad (10.41)$$

Now $F = (S_i/N_i)/(S_o/N_o)$ and with $N_i = k\,T_o\,\Delta f$,

$$F = \frac{S_i}{kT_o\Delta f} \cdot \frac{1}{b-1} \qquad (10.42)$$

Disadvantages are that in eqn. 10.42 the noise bandwidth Δf is not precisely known, and S_i must be a power value of the order $k\,T_o\,\Delta f$. Small power ratings of this kind are difficult to measure precisely. If one were to use as a signal generator a noise source with the noise power $k\,T_h\,\Delta f$ (where T_h would have to be known precisely), problems would not occur. This leads to what is known as Y-factor measurement, which is described in the following Section.

10.5.4 Y-factor procedure

The Y-factor method is by far the most widespread manual procedure for measuring noise factors. Figure 10.21 shows the basic measurement arrangement. The source resistance R_S is brought alternately to the two temperatures T_C (cold) and T_h (hot), and the quotient of the corresponding output powers N_{oc} and N_{oh} are determined, from which the noise factor can be calculated. With the cold source resistance R_S, there applies

$$N_{oc} = G \cdot k \cdot T_c \cdot \Delta f + N_a \tag{10.43}$$

With the hot source resistance,

$$N_{oc} = G \cdot k \cdot T_h \cdot \Delta f + N_a \tag{10.44}$$

An important point is the precondition that the same amplification applies in eqns. 10.43 and 10.44. For the Y-factor, one can derive the quotient

$$Y = \frac{N_{oh}}{N_{oc}}$$

The measurement is effected with the precision attenuator ATT, Figure 10.19.

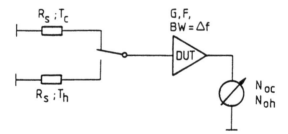

Figure 10.21 *Y-factor procedure*

From eqns. 10.43 and 10.44 it follows that

$$Y = \frac{G \cdot k \cdot T_h \cdot \Delta f + N_a}{G \cdot k \cdot T_c \cdot \Delta f + N_a} \tag{10.45}$$

from which

$$N_a = \frac{G \cdot k \cdot \Delta f \cdot (T_h - Y\, T_c)}{Y - 1} \tag{10.46}$$

Equation 10.18, however, rearranged is

$$F = 1 + \frac{N_a}{G \cdot k \cdot T_o \cdot \Delta f}$$

and with eqn. 10.46

$$F = \frac{\left(\dfrac{T_h}{T_o} - 1\right) - Y\left(\dfrac{T_c}{T_o} - 1\right)}{Y - 1} \tag{10.47}$$

and as a special case for $T_c = T_o$

$$F = \frac{\dfrac{T_h}{T_o} - 1}{Y - 1} \tag{10.48}$$

In eqn. 10.48, the value (dB) is in general designated as the excess noise ratio ENR:

$$ENR = 10 \log\left(\frac{T_h}{T_o} - 1\right) \tag{10.49}$$

Typical values for the ENR of measurement noise sources are 15–16 dB. The ENR value as a function of frequency is provided in the form of a calibration curve by the manufacturer for every noise generator.

Caution is required with the application of eqn. 10.48 because it applies to the special case $T_c = T_o$. In general, the room temperature is higher than 290 K, and is typically 295–300 K. Designating the noise factor derived from eqn. 10.48 as the 'measured value' F_m and the correct noise factor according to eqn. 10.47 as F_c, one can define a correction factor

$$dF = \frac{F_c}{F_m} \tag{10.50}$$

After taking the logarithm,

$$F_c \text{ (dB)} = F_m \text{ (dB)} + dF \text{ (dB)}$$

From eqns. 10.47 and 10.48 it follows directly that

$$df(dB) = 10 \log\left(1 - \frac{T_c - T_0}{t_0}\left(\frac{1}{\frac{T_h - T_0}{T_0}} + \frac{1}{F_m}\right)\right) \tag{10.51}$$

F_m is used numerically in this case. Figure 10.22 shows eqn. 10.51 evaluated for $T_c = 300$ K with the two typical values ENR = 5 and 15 dB as parameters.

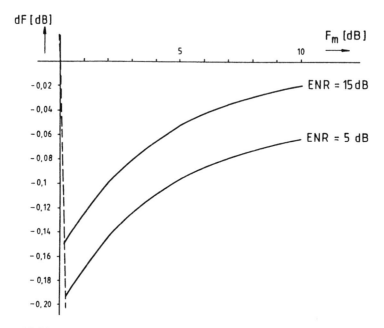

Figure 10.22 *Correction of measured noise factor for* $T_c = 300 K$

10.5.5 Automatic noise measuring stations

The manual Y-factor process described may be characterised by the greatest degree of precision, but is too time-consuming for many applications. In the case of series measurements, goods acceptance inspections, and comparison work, measurement stations are required, which immediately display the noise factor of the test specimen or changes in the noise factor.

There are two different kinds of measuring device on the market: automatic noise measuring devices, and automatic noise/amplification measuring devices.

Automatic noise measuring devices
With these units, measurement is carried out automatically by using the Y-factor process. The noise source is switched in a pulse pattern at, for example, 400 Hz

between T_c and T_h. The output power resulting from this, which jumps in 400 Hz pulses (eqns. 10.43 and 10.44 is processed in appropriate types of circuits in such a way that a direct display of the noise factor and the room temperature is provided. These devices do, in general, have disadvantages:

– in most cases they only have one single input frequency range.
– they adopt the standard value of 290 K for T_c.
– in most cases they only have low sensitivity.
– they measure the overall noise factor in accordance with the Friis formula (eqn. 10.28).

Automatic noise/amplification measuring devices

Modern measuring stations avoid these disadvantages by making use of a micro-processor and modern matching techniques. In addition to this, they contain a bandpass filter with a precisely defined noise bandwidth, and with the aid of this additional value determine the amplification of the DUT.

First, the inherent noise factor F_2 of the measuring station is measured in a calibration routine and stored. The basis for further measurements is the application of eqns. 10.43 and 10.44. Figure 10.23 shows the relationship in graphic form.

The device measures both the power values N_{oc} and N_{oh}, and initially calculates G and N_a from eqns. 10.43 and 10.44, with T_c and T_h being known, from which is derived, using eqns. 10.19 and 10.20:

$$N_a = G \cdot k \cdot T_e \cdot \Delta f \qquad \text{(as 10.19)}$$

$$F_{12} = 1 + \frac{T_{e12}}{T_o} \qquad \text{(as 10.20)}$$

the noise factor for the cascade DUT measuring station. From the Friis formula (eqns. 10.27 and 10.28), the system then calculates F_1 and T_{e1}.

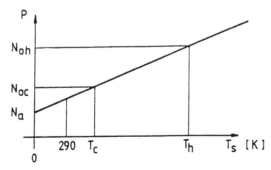

Figure 10.23 *Connection between source temperature T_s and noise output power P*

10.5.6 Sources of error in noise measurements

Noise measurement works with extraordinarily small levels, with the result that interference signals which pass to the input of the test specimen together with the measurement power will lead to erroneous measurements. There are, however, a large number of other possible errors, which are outlined briefly.

Electromagnetic interference

Causes: Interference, for example, from radio and television transmitters, or microprocessors operating in the vicinity

Remedy: Double-screened measurement cable, HF-proof housing, measuring cage required under certain circumstances.

Erroneous measurement due to unsuitable noise source

Causes: Measurement noise source has a different source reflection factor in the cold and hot state (Γ_c, Γ_h). When measuring a DUT with a small noise factor, even slight changes $\Gamma_c \rightarrow \Gamma_h$ can have catastrophic results. Example: Let F of the DUT be 1.10 dB. $\Gamma_h = 0.02 < 135°$, $\Gamma_c = 0.02 \leftarrow 45°$, measured value: 1.46 dB!

Remedy: One-way attenuator (isolator), or, if feasible, attenuator between noise source and DUT (which some manufacturers already incorporate in the noise source).

Nonlinearity

This includes a series of potential faults, the cause of which is frequently not easy to find; i.e. in these cases, experience and good measuring devices are essential.

Possible errors

— high error power, under certain circumstances outside the measurement frequency range, causes the measurement system to go into compression

— DUT and/or measurement arrangement are not yet at operating temperature

— drift among mains power units

— DUT fluctuates (probably outside the measuring frequency range)

— amplification of measurement arrangement too high.

Calibration problems

It must be guaranteed that the measurement noise source is calibrated perfectly. Likewise, the most stringent demands are also placed on the accuracy of the precision attenuator in the manual Y-method (Figure 10.19).

10.6 Appendix

10.6.1 Available power gain

The available power gain at a specific frequency is defined as the ratio of the available power P_{ao} at the network output to the available power P_{as} of the source:

$$G_a = \frac{P_{ao}}{P_{as}}$$

For a source with the output power $|b_s|^2$ and the reflection factor s,

$$P_{as} = \frac{|b_s|^2}{1 - |\Gamma_s|^2}$$

and

$$P_{ao} = \frac{|b_s|^2 \, |S_{21}|^2 (1 - |\Gamma|^2)}{|(1 - \Gamma_s S_{11})(1 - \Gamma_2 S_{22}) - \Gamma_s \, \Gamma_2 S_{12} S_{21}|^2}$$

with

$$\Gamma_2 = S_{22} + \frac{S_{12} S_{21} \, \Gamma_s}{1 - S_{11} \, \Gamma_s}$$

and therefore

$$G_a = |S_{21}|^2 \; \frac{1 - |\Gamma_s|^2}{|1 - \Gamma_s S_{11}|^2 (1 - |\Gamma_s|^2)}$$

One can see that G_a is a function of the network parameters and the source reflection factor Γ_s, and G_a is independent of the load reflection factor.

Index